Decorating with Marble, Stone and Granite

Decorating with Marble, Stone and Granite

CHRISTINE PARSONS

CHARTWELL
BOOKS, INC.

A QUINTET BOOK

Published by Chartwell Books
A Division of Book Sales, Inc.
110 Enterprise Avenue
Secaucus, New Jersey 07094

ISBN 1-55521-576-9

This book was designed and produced by
Quintet Publishing Limited
6 Blundell Street
London N7 9BH

Creative Director: Peter Bridgewater
Art Director: Ian Hunt
Designer: James Lawrence
Editor: Barbara Fuller

Typeset in Great Britain by
Central Southern Typesetters, Eastbourne
Manufactured in Hong Kong by
Regent Publishing Services Limited
Printed in Hong Kong by
Leefung-Asco Printers Limited

CONTENTS

INTRODUCTION

Smooth and hard, cool to your touch, there's a certain magic about marble, granite and stone that just can't be reproduced by any other material, and certainly not by any man–made or·synthetic imposter.

Stone has of course been in use for thousands of years: the Parthenon, the Temple of Apollo, and just about every castle and important ancient building has been made with some sort of stone, be it polished fine marble, rugged granite, or hewn from solid blocks of limestone or sandstone.

With this great heritage it's vital that restoration work and conservation continues to protect important buildings from the effects of twentieth-century living and to preserve great architecture for future generations. But it is not only the actual fabric of buildings which have been made from stone. Modern cladding techniques for walls and floors have long been favoured for interiors too, and it is the use of stone inside the home that this book is all about.

Firstly, the different qualities of each material are described – there is a huge difference between granite and marble for example. The different qualities make each type of stone suitable for different uses and applications. Look through the room-by-room sections which show ideas for all sorts of surfaces. Lastly and perhaps most importantly our catalogue of stones gives you an idea of just what is available. It is by no means complete as there are literally thousands of different colour marbles available from all over the world. What is available both in your locality and in your country by and large depends on the type of local stone and what individual importers are shipping into the country. But even so the choice is huge.

Remember though that stone is a natural product, that no two pieces will be alike even if they are cut from the same block, but that of course is just one of stone's intrinsic qualities: it's quite unique.

LEFT The simplest of colour schemes often works best, and black and white is always a winner. This floor is made from glossy, pale marble tiles.

CHAPTER 1

CHOOSING AND BUYING STONE

LEFT *Whether you are buying brand new units or simply looking for a way to give existing kitchen fitments a smart new image, this kitchen shows how real marble and a clever paint effect marry together perfectly. The beautifully thick worktops, floor and wall tiles have all been cut from the same softly veined grey and white marble, and the distinctive diagonal design has been echoed on the hand-marbled cupboard doors. A very clever use of real and fake marble that would work equally well in other colours too.*

Once you have decided to use some sort of stone in your home, you come to the task of choosing and installing it. If you are very keen and indeed fairly competent at home improvement you may want to fit it all yourself. A word of warning! While stone tiles and perhaps flooring in fairly small slabs is within most people's capabilities, I do not recommend that you try anything more adventurous. The jobs of laying, cutting and fitting stone are all best left for the professionals.

However there are a number of specialist companies who now deal exclusively with stone, so look in a local directory for firms in your area. Unless the job is huge, many of the big importers who say that they will 'supply and fit' will not however be interested in what they would consider to be small residential jobs. These large companies mainly deal with huge projects such as refurbishing office blocks, hotels and the like.

The other alternative is to go direct to an interior designer. This is an especially good idea if you have an inkling or two about the sort of style you want but need someone else to interpret it. A good designer will do this for you, adding some of their own inspirations too, to give life to your project.

Another good reason to deal with an interior designer is that he or she can, if you wish, supervise the project from beginning to end, from those initial drawings to selecting the stone, and finally to organizing and supervising all the work.

To get the exact colour stone you require, take a trip to the importer or supplier's showroom, either by yourself or better, with your designer. This is important as it is here where you will see exactly what is in stock or could be ordered for you and delivered in a reasonable time. It is certainly always worth considering locally mined stones as they are likely to be in greater abundance and there is some kudos in putting your money in a 'home industry'. However any importer worth his salt should be able to show you a good selection of different stones from various parts of the world. What is vital is that you find out if your supplier has sufficient stone to do the job you want. Always check every piece you are buying to see how the colour, grain, veining etc matches. You may have to work out how best to position them to achieve an harmonious balance between slightly differing sections.

IN THE WORKSHOP

Once the stone has been transported from the quarry to the workshop, it is ready to be cut up into small pieces. Slabs of different sizes are the most commonly used components so each block must be sawn into many pieces.

Saws

Wire saws may be singular or multi 'bladed', but most cut through the stone with the help of impregnated industrial diamonds.

A CIRCULAR SAW makes just one cut at a time. The stone block is mounted on an operating table and the saw sits inside a frame. Fed with water and abrasive to keep the blade cool while in use, this type of saw is used for squaring off irregularly shaped blocks, to cut a single block into slabs of different thicknesses, and for cutting slabs thicker than l5cm (6 inches).

A FRAME SAW may have anything from 10 to 25 diamond-tipped blades, all moving backwards and forwards under tension at a controlled speed which differs depending on the type of stone being cut. It's slower to use than a circular saw but with a longer blade length it can cut wider blocks, and of course it is making a series of cuts, not just one. It is used for making slabs of less than l5cm (6 inches) thickness and for greater accuracy.

In a granite yard, there may be a frame saw with teeth for cutting, used with either silicon carbide grit or fine steel shot. This is necessary because granite is such a hard stone, and a diamond-tipped saw would soon become blunt. Some sandstones which require a finely scored surface finish are cut with a corrugated reciprocating blade, used with steel shot fed in water. This leaves a distinctive gouged surface.

This first sawing process is called 'primary' sawing, leaving slabs sawn on two sides. Secondary sawing involves a rotating table and two independently adjusted circular saws to cut the other four sides of the slabs, while some of the most modern saws operate by lasers. The completed slab, sawn on six sides is now ready for use, or it can be further machined or hand worked.

MACHINES have been designed to saw or grind the stone to give an angled cut, cut slots and grooves and for drilling. Planers are used for continuous runs of complicated mouldings while a router can cut patterns and mortise joints. Automatic polishing machines give marble and limestones their smooth, high gloss finish.

HANDWORK The skill of the stonemason comes to the fore in the finishing of stone. Although much can be achieved by machine, he can obtain a smooth finish using a chopping hammer or cold chisel, or create roughness by working across the surface of the stone with a pointed pick or punch. Using some mechanized hand tools, a skilled stonemason is still in high demand, for his craftsmanship in carving, restoration and for the specialist jobs where no machine can or has not yet been designed to do the job.

Importers carry their stone stock in various forms:

BLOCKS of stone, waiting to be cut to a customer's specific requirements.

SLABS usually l9mm (¾ inch) thick for general use on worktops, floors and walls but also available in other sizes. The length and width of individual slabs is governed by the size of the block they were cut from. Suffice to say there is a good choice - with sizes up to 1.5 - 2m (5–6½ feet). You can tell if any individual slab is sound by 'ringing' it like a crystal glass. When knocked with a hammer or piece of stone it should ring clearly, but if the stone has a fault, the sound will be dull.

TILES Ideal for both walling and flooring, stone tiles are practical and easy to fit in more or less the same way as standard ceramic floor or wall tiles. Some tiles will be made from an agglomerate mixture: chippings of marble or stone bound together with special resins. Manufacturers claim that they are a better bet than real stone as they tend to be thinner and are therefore lighter, and also because they lack both porosity and cracks. Whether they are better for your particular purpose is your choice but, in the end, the look you want to create and, most important, your budget will be the deciding factors.

PANELS It is now possible to buy thinly cut marble which has been fixed to a special backing making it much lighter and more suitable for large expanses of wall. The marble itself is about 6.5 -7mm (¼ inch) thick and various manufacturers sell it in large sheets some 12 by 24m (4 by 8 feet) or in small tiles sizes. Whichever you choose, the material can be riveted, screwed or glued to the support structure; it can be cut using a jet water cutter to fit around switch plates and lights, and the edges can be shaped in the same way as ordinary stone (see below). Because the stone can be cut so thin,

some companies are even offering strips of different colours on both the sheet and tile so a striped or banded effect can be created. As this material is so light it can be fixed direct to breeze blocks, studding partitioning, brickwork and even to ceilings.

FINISHES

The way the edge of the stone is finished off can make or break the final effect. If the interior designer has done his job properly there will be little finishing to do 'on site', with perhaps just a grinding and polishing machine to finish off a few edges. If large slabs are being installed the designer should have accurately worked out how they will fit and have had them pre-cut either by the the supplier or at his own workshop.

Edges can be cut and finished in a number of ways:

BEVELLED edges have a slope.

ROUNDED edges have had the corner ground off to give a smooth arc on the top edge or in the case of a slab, it can be on two edges

SHAPES can be cut in the stone, either by hand or by machine to give contours - you find them on fire surrounds for instance.

SPLASHBACKS AND SKIRTINGS are cut from a slab and placed as an upstand on either a worksurface or floor.

RESTORATION AND CLEANING

Like any other material, stone gets its fair share of day to day wear and tear, a build-up of dirt, and of course occasional accidents do happen, so swift action is needed to clean up spills.

CLEANING MARBLE

Marble is extremely porous. Although it is hard to the touch it absorbs liquid easily - and to a great depth – so stains can be hard to shift. So the first and most important rule is never use a valuable or prized piece of marble as a work surface or coffee table – that would be just asking for trouble. The second rule is to clean up stains as quickly as possible, before the liquid has had time to penetrate too far into the surface. Some of the following cleaning methods involve the use of strong stain removers, so if you are in any doubt about using them, or spoiling your marble, the third and final rule is don't! Instead seek the help of an expert restoration and cleaning company.

General cleaning

To get rid of light dirt, most marble responds well to this simple cleaning process: fill a bowl or bucket with warm water, then add a cup of pure soap (never detergent). Place the marble on a flat surface (small objects can be submerged in the water for a short time), then using a medium-hard brush, work up a lather and scrub across the surface. It's important that you don't leave the lather on the marble, so as you work, rinse it off with fresh, clean water, and wipe the surface with white rags or white paper towels.(Please don't use coloured towels or paper as the dye may come off onto the marble.) If the marble is really grubby - from a sooty fireplace perhaps - add a few drops of ammonia to the soapy water.

Polishing marble

Undamaged marble which is perfectly clean can be polished with a silicone furniture polish or either a brand-name or home-made beeswax polish. They will all give the marble a soft attractive sheen and a little protection against water. The polish is best applied in several light coats rather than one thick one. To start, rub the surface over with a clean cloth to remove any light dust or fluff, apply the polish and buff over with a soft cloth. On large flat surfaces you can use a mechanical polishing machine but take great care not to scratch or scour the surface of the marble.

The traditional way to polish marble and remove light stains is with a mild abrasive. You can use powdered chalk, pumice powder or tin oxide, rubbed on to the surface on a damp white cotton cloth or a chamois leather. Work it in by rubbing the powder on to the marble in small circles, moving across the surface. When it feels smooth, rinse the powder off and wipe dry with white rags or paper towels.

Stain removal

If the use of a mild abrasive doesn't remove a stain, try one of these methods.

GREASE STAINS Caused by a greasy, fatty or oily liquid seeping into the surface but not actually damaging the pores of the marble, the simplest way to draw it out is by using a poultice. If possible lay the marble on a flat surface, then mix a small amount of paste made from using equal quantities of kaolin and benzene or potato flour starch and white spirit. You can make the paste extra effective by adding a few drops of solvent such as acetone or ammonia. Apply the poultice to the stained area and cover it with a piece of plastic or aluminium foil to prevent it drying out too quickly. Leave it for about 30 minutes, then remove the cover and allow the poultice to dry. Rinse away the residue with a damp cloth or chamois leather and check the stain. You may have to re-apply a freshly mixed poultice several times before the stain disappears.

STUBBORN STAINS To remove stubborn stains, a stronger stain remover must be used. Make up a mixture of one-third 20-volume hydrogen peroxide and two-thirds water. You will also need to have a syringe or old eye-dropper ready, loaded with ammonia. With the marble positioned so it is perfectly horizontal (prop it up with lumps of plasticine if necessary), pour just one teaspoonful of the peroxide/water liquid over the stain, then immediately add a few drops of ammonia on top. The peroxide solution will now start to bubble, so leave it until the bubbling stops, then rinse the area straight away with plenty of clean water to make sure you remove every trace of bleach. You can repeat this whole process if the stain persists. For stubborn stains on vertical surfaces, make a poultice by moistening a clean white cotton rag with the peroxide solution and tape it on to the marble using acid-free masking tape. Check the stain regularly and rinse well as before.

Marble floors

In general, never use harsh abrasive cleaners on marble floors as they will take away the shine and could eventually discolour the surface. However, simply because floors do suffer so much wear, the use of some sort of detergent is almost unavoidable. Damp mop the floor in sections so that the detergent solution is never left for too long, scrubbing any stubborn marks where necessary. Remove the dirty water as soon as possible by rinsing with fresh clean water. To bring up a shine use a self-polishing silicon wax floor polish, or apply a transparent seal. There are specialist cleaning companies who will do this for you.

Mending marble

It's almost inevitable that at some time or another your marble surface will suffer some surface damage from scratches or knocks. Small sections of marble can be glued back in place using a strong adhesive such as a two-part epoxy. Follow the maker's instructions to the letter, and if necessary either hold the damaged section in place using a heavy book, or clamp the pieces together while the join hardens. Do not be tempted to remove the weights or restraints too soon. Epoxy adhesives need to be left for up to a week before they reach their maximum efficiency.

Getting in the professionals

They are some restoration jobs that really should be left to an expert. It can for instance be extremely difficult to mend chips and grooves in marble, or indeed bad scratches, and specialist companies have the hardware and the know-how to restore the surface perfectly. They make full use of resins and fillers to disguise missing sections, and have efficient machines that can regrind the surface to perfection.

BELOW With such strong architectural lines, huge picture windows and stunning plants, only the simplest colour scheme was needed in this garden room.

CHAPTER 3

KITCHENS

Stone in the kitchen? Somehow the very idea conjures up thoughts of a dark, below-stairs Victorian affair with parlour maids labouring at those now sought-after stone butler's sinks, or down on their knees scrubbing flagstone floors. It's not really the sort of kitchen environment that any of us considers our 'dream', and it's certainly a world away from the modern pristine, wipe-clean, germ-free kitchen we've all come to know and love. But just as kitchen design-

ers have returned to the use of natural wood for units, it's only in the past few years that marble – and especially granite – has become a highly desirable surface for kitchen worktops.

Of course marble is renowned for its cold-to-touch surface that makes it perfect for pastry-making, but more and more designers are now favouring it not only for worktops but to create an undeniably luxurious effect for floors and walls too.

LEFT In this large town-house kitchen, the owners wanted to make the most of the space with a light open effect, but this solid bank of plain white units gives an almost puritanical feel. To offset the block of neutral colour the softest grey and white marble has been used extensively, not only for work surfaces but also for the floor and walls between base and wall cupboards. The designer has chosen large sized tiles for the floor, with slabs for the walls and surface. To ensure continuity of effect, as here, it's important that all the different pieces are cut from the same block of stone.

ABOVE British furniture designer Mark Wilkinson specialises in fitted kitchens and bathrooms, so when it came to building his own kitchen he naturally went for the best materials available and chose extravagantly thick marble

slabs for all the work surfaces. A softly veined cream marble works beautifully to show off the delicate frosted oak base units. As a distinctive finishing touch, the edges of all the marble have been hand carved with a simple border design.

ABOVE Great thought must be put into any kitchen design if it is to be an organized, efficient and pleasant place both to work and inspire the culinary 'juices', so in this square room a central preparation area not only looks stunning but makes practical sense. Not only has the well-trodden path that forms the working triangle of sink, preparation and storage been kept to a minimum but the surface itself, made from granite, is tough enough to use for just about anything from topping and tailing vegetables to baking, pickling – or even as an impromptu table for breakfast or a coffee break.

RIGHT With the right choice of materials stone can be as plain or as colourful as you choose it to be. In this kitchen two strongly coloured but quite different materials work well together. On the walls there are quartzite tiles from China, a really colourful range of gingery browns and greens with tiny patches of blue and yellow, while the floor is covered with African slate tiles with a blue base and heavy markings of gold and brown.

LEFT Give a kitchen a warm, earthy flavour by choosing natural materials wherever possible. This farmhouse-style room has a cosy feel that has been achieved with the use of wooden units, together with a polished stripped wooden flooring and rose-coloured granite work surfaces.

WORK SURFACES

Of all the various types of stone available there are only two that are really suitable for kitchen work surfaces, and indeed one is even better suited than the other. They are, namely marble and granite. Marble has of course been used for many a year as a pastry board. Either inset into another surface – wood or melamine faced chipboard – or as a portable board, you can't beat that wonderfully cold surface to keep your flour and fat at the perfect working condition. However in recent years marble has found great favour in a wider application and is being used throughout kitchens. It certainly looks stunning, but the owner of marble work surfaces has a great responsibility to keep them looking in tip-top condition. He or she will have to think ahead and be prepared to do any sort of chopping and cutting on a wooden or laminate chopping board instead. The reasoning behind this is simple: marble scratches. Although it feels hard, it is a fairly soft stone and you can be sure that every knife mark will show, but perhaps worse is the fact that

RIGHT Joins between slabs should be positioned where they will show least. Straight joins at corners look best. These thick brown granite slabs have a simple but effective finish: a straight groove has been cut along the front edge.

these scratches will soon discolour and look really unsightly.

Granite is therefore a much better bet for kitchen work surfaces. This heavy-duty stone has been used extensively for exteriors of buildings so its toughness record is legendary. This, of course, makes it a perfect choice for the kitchen where over a period of time it may literally be hit with everything from chemicals to acids. Just think of all the things that get cut up, spilt and wiped on and off: fruit juice, vegetables, grease and fat not to mention all sorts of cleaners and bleach. You can rest assured that granite can cope with them all: its glossy polished surface is virtually indestructible.

Fitting work surfaces

Considering the huge weight of the stone itself, it is surprising that the quality of your base units doesn't matter too much as the weight of the heavy stone is spread so evenly across them. So happily even the cheapest home-improvement units are as suitable as the most expensive handmade cupboards.

However, to give a solid, level surface on which to lay the stone, most fitters will first position and screw sheets of 20mm (¾ inch) thick MDF (medium density fibreboard) on top of the units. The marble or granite is then laid above and the levels are checked. This is because both marble and granite can vary in thickness from one slab to another. Even though the sawing process is completed by machine, there can be a very slight difference of a few millimetres. In fact some slabs can even vary in thickness from one end to the other. Once the levels are checked, the stone is bedded down either in a tile cement or with a cement screed and any irregularities in the thickness that do

show on the front edge can simply be ground down to match. To disguise the chipboard below the stone, an extra front piece is added. This narrow strip of stone not only hides the board but makes the stone look twice as thick – a very clever visual trick at very little extra cost. This front piece is usually set back slightly which also gives a neat detailed effect.

Finishing off

The front edge of the work surface can be finished off in a number of ways. Square edges are simplest but the stone could also be rounded on one or two sides, or a carved detail or groove can be added on a border.

For neatness and cleanliness a simple backstand or splashback can be added at the back of the work surface. This should be cut from the same piece of stone as the surface itself to ensure continuity of colour and pattern. It can be any height you like, though somewhere between 5 and 20cm (2 and 8 inches) gives the most pleasing proportions. Again the top edge of the backstand can either be left as a squared off edge or the front edge can be rounded off.

BELOW Sinks can be fitted into stone work surfaces in one of two ways: either 'inset' or 'on-top'. Whichever method you choose, the hole is cut in the same way using a template. An on-top sink simply sits in the hole, hiding the imperfectly cut edges. For these inset sinks, the edge of the hole must be ground and polished to match the surface, and in this case the bowls are fixed on to the back of the slab.

LEFT Stone blends well with other materials to give versatility in a busy kitchen. This grey and white granite has been used alongside wooden work surfaces topping a bank of cupboards to create a peninsular. It gives a practical area that can be used both for preparing food or as an elegant surface for informal entertaining.

RIGHT Set lower than the surfaces used for chopping, this granite slab is the perfect height for pastry making. The back of the fitment has been cut to take the slab giving an interesting side detail.

FLOORS

What qualities does a kitchen floor need? It's got to be practical, easy to clean, tough and hard-wearing, so stone – which is all of these things – becomes a natural choice. Even better, it will last a lifetime and grow ever more beautiful as the years pass. Old flagstones and tiles, lifted from buildings before they are demolished, are highly sought after and expensive to acquire but they do give an instant 'old' effect, even in a brand-new kitchen.

Getting down to basics

Thorough preparation of your sub-floor is absolutely vital if your stone is to lie perfectly even. A concrete sub-floor is best as it is both solid and strong, and if it is somewhat uneven it can easily be levelled. A wooden floor on the other hand poses the most problems. Not only is it likely to be uneven, but it has a tendency to move, especially if your home is centrally heated and the house is constantly warming up and cooling down.

WOODEN FLOORS To give a wooden floor the strength and rigidity needed to take the weight of stone, a layer of plywood should be fixed on top. But before this is undertaken, you must make sure that the ventilation under the floorboards is adequate to prevent any problems with damp or dry rot. Remember too that once the stone floor is in place you will not have any access to underfloor pipes or cables which could cause problems in the future. The plywood should be stacked in the room in which it is to be used for 48 hours before fitting so it adjusts to the atmosphere. Then it should be laid, brick-fashion with staggered joints, screwed into place across every panel at 15cm (6 inch) spacings to ensure a solid bond to the floorboards. As an alternative to this plywood method some builders will take up the floorboards completely and instead of plywood will lay a new floor using a flooring grade chipboard.

CONCRETE FLOORS To level a concrete floor use a self-levelling compound. This is supplied as a powder which you mix with water. Before you start the floor must be clean and free from damp. Pour some of the compound in to the furthest corner from the door and spread it with a trowel until it is about 5mm (¼ inch) thick, then leave it to find its own level. Continue working across the floor until the whole floor is complete, finishing at the doorway. This compound can be walked on after about an hour, but it is best to leave it for at least a day –

LEFT Slate tiles give an unusual, rustic atmosphere to a kitchen-cum-dining room. These are African slate, cut with a hand-held guillotine which gives them distinctive chipped edges. The tiles have two faces: a larger face with the best concentration of colour or a smaller face which gives a cobblestone effect. Here the tiles show their larger face with the soft blue background and orange, red and gold markings.

RIGHT You can create a traditional flagstone effect by using a combination of different coloured stone. This floor lends its subtle patchwork effect to a mixture of tan, green and yellow tiles cut from Brazilian quartzite. The charming lived-in look comes from the use of random tile lengths cut in a variety of different widths.

and preferably a few days to harden completely.

Laying the stone

Tiles and slabs are laid in a similar way, either into a flooring grade tile cement or in a cement screed. To ensure the tiles are laid in perfectly even rows, two wooden battens are fixed in place to the floor at right angles to each other to act as guidelines. The first line of tiles is laid against one batten, while a third batten is used to hold each line of tiles parallel as it is laid. The third batten is moved down the floor as each row of tiles is completed. Once all the whole tiles have been laid, others can be cut to fit around the edges. Slabs however need more organization. Your architect or builder should have these pre-cut to fit your floor exactly before they are delivered to your home. They will be numbered with a key for easy installation.

WALLS

Most kitchens do not have much free wall space. It is usually taken up with cupboards, so huge expanses of marble or stone are not necessary, and the most effective use is to clad the gap between the base units and the wall cupboards. This looks particularly luxurious if you use the same material for work surfaces. Marble or stone tiles are simplest to fit in exactly the same way as ceramic tiles. Thinner than slabs, they are easier to cut and fit around awkward shapes such as electric sockets and light switches. They can also be fixed using a brand-name tile adhesive or a cement-based adhesive. Slabs of course are heavier, so need a firmer fixing. They are usually hung on the wall with wire fixings while the back of the slab is cemented into place (see 'Bathrooms' page 33).

RIGHT In this designer kitchen, the marble slabs have been thoughtfully worked out to incorporate a recess for a towel-style radiator. The marble has also been carefully cut to fit around the wall cupboards – not behind them – which would be a dreadful waste of good marble – and money.

RIGHT Every piece of fossilized stone is quite different so these tiles make an unusual and interesting choice for a wall surface with their infinite variety of colours and pattern. Co-ordinating work surfaces have been cut to fit into a corner, while thin front pieces have been fixed below the work surface to hide the chipboard base.

TOP RIGHT Lavishly used on both walls and work surface, this soft grey granite looks stunning and it is the perfect foil for a strong modern kitchen.

BELOW Granite tiles, cut from the same rock as the slabs used for the work surfaces, create a wipe-clean splashback behind a sink. Without the benefit of natural light from a window, full use has been made of subtle under-cupboard strip lighting instead. Easy to install, these neat lamps highlight the beautiful grain and colour of the granite.

LEFT Hand cut slate tiles, with their uneven, rough surface make a sharp and highly effective contrast in this sleek, all-wood kitchen. The tiles are used just to the height of the first shelf, with bricks above chosen to match the red-brown tones of the slate.

BATHROOMS

Of all the rooms in a modern house, marble is used most in the bathroom. Think of the sets in the popular TV soaps and the bathrooms are sure to have marble in them somewhere. There's no denying it, it is a luxury material, though it doesn't have to be used on every single surface to work its magic. Use marble to surround a bath or use it as work tops for inset vanity basins, teamed with either plain white or toning sanitaryware. A marble floor or marble tiled walls is also the perfect foil for a coloured bathroom suite, or it can work well the other way around: with the suite made from reconstituted marble, and plain walls and floors.

BATHROOM PLANNING

The positioning of your bath, WC, bidet and hand basins depends mainly on your plumbing. While it is a relatively simple and inexpensive job to re-site a water supply pipe, it can be extremely costly to start moving waste or soil pipes around, so if you can, work existing pipework into a new layout. This of course means that your WC will more or less have to stay put, but other items can be shifted around to suit the room and your own personal needs and wishes.

Working out a new layout needn't be difficult. The simplest way is to draw a scale floor plan of your bathroom on to squared paper. From another sheet of squared paper, draw up each piece of sanitaryware and cut out the shapes. Use them like a jigsaw, moving them around on your floor plan until you have a scheme that works. One word of warning! Don't place items too close together – remember that you have to leave enough room to hop in and out of the bath, to stand in front of the wash basin and to use your WC and bidet.

Now decide where and how you want to use your marble or granite. In bathrooms, marble is the most commonly used stone, but granite would make an excellent alternative for a floor or vanity unit top.

ESSENTIALS

Storage

The smaller your bathroom the more important storage becomes, so it's worth allowing extra space for a cupboard or shelves. If space is very limited, why not box in a pedestal basin to create a cupboard – useful space for all those items you'd rather keep out of sight, or build in a 'secret' cupboard under one end of the bath. For favourite bottles and everyday items, a set of glass shelves will almost disappear into a wall or you could build them in to an alcove.

Heating

A marble bathroom is likely to be on the chilly side, especially during the winter months, so some form of heating is a must. A heated towel rail or a towel/radiator will solve warmth and storage problems in one go. For no-show heating, choose a skirting radiator. These are very long but thin and sit almost at floor level, or you could turn one on its side and position it vertically in a corner.

LEFT Marble can be used most successfully in small amounts. Here steps have been built up to the rim of this bath to give it a sunken look. More real marble has been used as a vanity top, while a clever paint technique has been used to paint 'marble' on to wooden surfaces to match.

Ventilation

When warm air hits a cold surface condensation is formed, so along with good heating, some sort of extractor fan will also be needed in a bathroom fitted with marble or stone. This extractor can be positioned in a wall, in a window or ducted up through a ceiling. The cheapest types are hand operated with a pull-string but more efficient is an electric extractor that can be activated manually or with a sensor that automatically switches on when a certain moisture level is reached.

Lighting

Your choice and the placing of light fittings is especially important if you are having marble clad walls, as electrical cables and connections must be made before it goes up. Decide on the mood of lighting you want: down-lighters for instance throw soft pools of light and when worked on different switches can highlight just one area of the bathroom at a time. If you are planning to use your bathroom for make-up, good light around the mirror is essential. The lamps should be placed so the light is thrown directly on to your face.

RIGHT Simple in design and beautifully made, this basin stand from British furniture designer John Lewis of Hungerford features a relatively plain marble top and a shaped upstand, set on a wooden base which is in turn supported on pewter legs. Simply beautiful.

SINKS AND BASINS

It is possible to buy basins made from marble, and certainly in Italy and especially Greece you will find beautiful marble sinks and basins in even the poorest homes. However you are more likely to come across a reconstituted marble basin, either inset into a top made from the same material, or to be used in a melamine-faced chipboard top or indeed inset into real marble work top. But because it's a simpler option, many people plump instead for a ceramic or moulded plastic bowl that sits either on top of a marble or granite work top or one that is fixed below it.

Just as for kitchen work surfaces, all the same rules apply (see page 20) with the base cupboards or vanity unit itself supporting the weight of the stone, with a piece of chipboard placed on top if necessary, to give a completely flat surface.

Where bathroom work tops do differ from kitchen surfaces is in their shape. Most vanity work tops in bathrooms are curved or shaped with a pretty rounded edge. Where two basins are to be set in a single worktop, the curves can go in and out to emphasize this. Holes are cut into the stone to accommodate the bowls and if you choose the type of bowls that sit on top of the holes, the taps are usually set into the basin itself. If however, you choose basins that are fixed below the line of the stone, holes will need to be cut in the stone to accept the taps as well. Because of this you should decide on the style of tap before the marble is fitted. As taps vary enormously in size from different companies once the holes have been cut you may find that different taps will not fit. *Mono-bloc* taps are single taps – one for the cold, one

RIGHT A slab of almost white marble with a highly polished gloss surface has been used to top a vanity unit in this tiny town-house bathroom. Cut to echo the lines of the shaped floor cupboard, both front edges of the marble have been ground to give a soft rounded edge.

for the hot water. A *mixer tap* will have a single hole fitting but with one tap and two knobs to control the temperature of the water. A *hand/ shower mixer* is usually set over a bath, but if could be worth fitting if you use a hand basin to wash your hair.

If your stone surface is not to have any cupboards underneath it will need some strong supports such as wooden or metal posts instead.

BATHS AND SURROUNDS

You generally have the choice of either a marble bath or a marble sur-round. Now if you opt for the former, your bath is almost certainly going to be made from a reconstituted marble, moulded into shape. These do however look very realistic, and you can have a whole suite made to match. For total co-ordination match-ing panels can be made to clad the walls too. If on the other hand you would rather opt for a traditional bath – made from cast iron or one of the modern moulded plastics – you can simply set it in a marble or granite surround.

The bath itself will need to be firmly anchored inside a wooden framework which must of course be strong enough to support the weight of the stone surround. This frame should be covered with a marine plywood (the waterproof type) or a chipboard. To give any bath a sunken style, if the structure of your home allows it you can drop the level of the bath below the floor. However in most houses this is not possible so the alternative is to raise the floor instead. This is usually done by building one or more steps up to the bath, which of course can all be clad in stone.

Where space allows, a corner bath or an unusual circular, oval or even octagonal shape will give your bath-room an air or opulence, but on the practical side, you must make sure that your hot water boiler will be able to supply enough water to fill it! Above all else your bath and its sur-round must be waterproof so the seal between the two must be excellent. Make sure that your builder uses a suitable flexible sealer, that will 'give' a little. After all it will have to be able to cope with both the fluctua-tions of temperature and the weight of the water and indeed a person get-ting in and out of it too.

BELOW This alcove has been completely lined with marble, with recesses cut for a mirror, shelf and soap. This shows the importance of planning and design, taking into consideration the pattern, colour and variation of the marble.

RIGHT If you have a large bathroom, give your bath centre stage. This unusual bath is large enough to accommodate four people, and to give this rich, rather masculine atmosphere, a dark brown and beige granite was chosen for the whole of the floor area. Note the clever 'fakes' too, from the 'marble' trim around the edge of the bath to the plastercast Greek columns and painted dado, all decorated with a clever 'marble' paint effect.

BELOW *Black is back. This tiny bathroom needed a really strong scheme to show off a plain white bath and this black speckled marble does the job in style. Strips of marble have been cut and laid around the edges of the bath, with another piece cut to fit the windowsill, and the whole lot matches the floor. Simple but extremely effective.*

RIGHT *This stunning mock tortoiseshell bath needed a sympathetic colour scheme that would help to set it off yet still have an impact of its own. The answer was quite simply to set the bath in a beige toned marble, with walls and floors to match. To pick up the colour, the designer has chosen a matching dark marble to use as an inset, outlining the shape of the bath and continuing it up the walls to highlight an over-the-bath mirror.*

LEFT *Rhapsody in blue. If you have a family of small children, or simply tend to splash water around a lot, then it's a practical idea to continue the marble from a bath surround up the walls too. The marble will simply shrug off water, and so long as the marble has been laid in a suitable cement and sealed properly, the whole area will remain watertight.*

FLOORS

Bathrooms are usually up on the first floor of a house, close to the bed-rooms, so they are unlikely to have solid concrete floors. As any move-ment from floorboards is likely to cause the stone above it to move, you must prepare your sub-floor before the marble is laid. The same basic ground rules apply as for 'Kitchen Floors', (see page 24), except that in a bathroom it's vital that a marine plywood is used to line the floor. This wood is used in the building of boat hulls so it's guaranteed completely waterproof – essential in a bathroom.

Once the floor has been thoroughly prepared it is ready for the stone or marble slabs or tiles. If you are using marble floor tiles, the most com-monly used size is 30cm square,

which is more or less equivalent to the imperial 12 square inches. A well planned floor should have equal sized tiles at the edges and the corners, and to achieve this tiling must begin in the centre of the room. If you simply start in one corner and work out-wards you may end up with tiny sliv-ers of tile down one side and huge chunks down another.

To find the centre of your room first find the middle point on each wall. Then, stretch a length of chalked string across the floor on opposite walls and snap it to leave a mark in the shape of a cross. Where the two lines intersect is the central starting point. Start to lay the tiles by working on just one quarter of the cross at a time, working out-wards in both directions to form a right angle.

BELOW Beautiful old flagstones in random sizes form this lovely patchwork effect floor. Because of its irregular shape, many stones had to be cut to fit on site using a special stone cutting saw.

RIGHT Agglomerate tiles are made using chips of marble and stone set in a resin which not only gives a high gloss finish, but also ensures an even distribution of colour and shading throughout every tile. A good choice for a bathroom as this type of tile is lighter than solid marble, and excellent for a large area, the tiles come in a standard 30cm (12 inch) square size, but special sizes, treads, risers and even anti-slip inserts can be made.

WALLS

Marble tiles are, of course, applied to walls in just the same way as ceramic tiles, and indeed they can be fixed using a ceramic tile adhesive, though in areas of constant splashing a waterproof type must be used. Alternatively they can be fixed using a cement based adhesive. Stone and marble tiles do however differ in one important way and that is that they have perfectly straight edges so they butt up together more closely than ceramic tiles do. Just as the gaps between ceramic tiles are filled with a grout, these tiny gaps between marble tiles must also be filled to give a waterproof finish. However any good company will mix up a special

BELOW For an extension to a period house, an instant aged look can be created by building just one stone 'feature' wall.

filler, coloured to match the base colour of the marble so that once in place, it hardly shows.

Just as a floor needs preparation, so do walls. Older houses in particular need special attention. Traditionally built lathe and plaster walls will not take the weight of either heavy tiles or slabs so it may be necessary to line walls completely with plywood just as for floors.

Slabs which are of course very heavy are fixed to the wall twice. First holes are drilled at the top of each slab to take rigid 'wires' which literally hang the slabs from the wall like a picture. As added security the back of each slab is also bedded into a cement mixture. Where slabs butt up to each other, the tiny gaps are filled with a coloured epoxy resin, just as for tiles.

ABOVE Lightly veined white marble is elegant enough to stand alone, here lining the walls and creating a sumptuous effect.

BELOW *The bathroom is perhaps the only room where you can use marble so lavishly, here creating a warm effect.*

SHOWERS

Showers are subject to more wetting than any other part of the bathroom, so it's here in particular that attention should be paid to creating a water-tight finish. The marble or stone finish is only as strong and sound as the surface to which it is being fixed, so if for instance a cubicle is to be built, it must be made both rigid and sturdy so that once the tiles are fixed, it will not move. Any movement will cause the tiles to shift and allow water to seep beneath them.

RIGHT *Marble comes in a vast range of colours and patterns. This pink and grey design is complemented by an all-white suite.*

CHAPTER 5

LIVING ROOMS AND HALLS

When it comes to the living areas of the home, stone is mainly used for fireplaces, for floors and sometimes for furniture. But with imagination and flair it can look either homely or lavish, traditional or ultra-modern – in fact just about any style can be created.

Stone can be used architecturally to excellent effect on interiors as well as on the outside of buildings. Columns and pillars, mouldings and cornices can all be carved from stone.

LEFT If you have a large floor area to cover, make the most of it with an intricately designed floor. It will look splendid! This one features large cream coloured travertine tiles set inside bands of white and grey flecked marble with black marble squares at the corners. The room, designed by David Hicks International, has three identical windows which have been highlighted with arch-topped sections of travertine, then finished off with shaped stone reveals.

BELOW This large, light, generously proportioned living room owes most of its design to stone. For the floor, large marble floor tiles have been inset with small black marble squares, while specially made stone columns support a deep stone frieze which runs completely around the room. Finally, two beautiful carved marble lions form the base for a thick glass table top.

LEFT *If you're a collector of anything – be it pictures or pine cones – your room should have a fairly simple colour scheme to show off your treasures. This room has a stylish ceramic tiled floor, set off with an unusual stone table and plinth. The warm cream walls are a good foil for the beautiful stone door surrounds and shaped cornice sections.*

FIREPLACES

Stone is certainly at home in the fireplace, being both practical and attractive. The surround can be made from a great number of stones – both natural and reconstructed. Marble is a favourite both for the surround itself and to form a decorative hearth and slips – the flat surfaces between the edge of the fire and the surround. Slate is also used, as is granite, onyx, quartzite and limestone.

The style and design of a fireplace should of course echo the period of your home. Some people are lucky enough to buy an old house complete with original fireplaces, others may be discovered under boards, or even bricked up openings. However if your home has no fireplace at all, you could fake it by fitting a surround against a flush wall. It will give a square or boxy room a focal point, or you could go the whole hog and have a chimney/flue installed to create a working fire.

Having any stone fireplace replaced or installed is a dirty, heavy and skilled job, and therefore best left to a professional. Thankfully there are many companies who specialize in

RIGHT Perfect for a large room, this stone fireplace is a one-off design featuring hand carved motifs, a shaped opening and a splendid galleon centre piece.

RIGHT African slate has been used to clad the lower half of this entire chimney breast for a country-style fireplace. Most slates from Africa are Silver Blue – a lovely two-tone effect created by a dark blue with a silver overlay.

50

designing and installing fires and you can call upon them to fit the fire of your dreams.

Care of your fireplace

If your fireplace is in regular use, dirt and grime will build up, so regular cleaning is needed to keep the stone looking good. Avoid standing cups, glasses or ashtrays on it and be particularly careful not to spill liquids such as wine, tea or coffee over the stone, as marble in particular is extremely susceptible to staining.

Always let the hearth and surround cool down before you clean them.

Clean *stone* by sponging off marks with warm water, adding a little detergent if necessary. For stubborn stains, use a stiff brush.

On polished *slate* use a stiff brush and warm water mixed with a little detergent. On unpolished slate an abrasive cleaner can shift stains.

Marble and granite should be washed regularly with warm soapy water, then dried and polished up using a chamois leather. For a gloss finish the stone may also be polished with a wax polish, applied sparingly, or a special stone polish.

BELOW Totally classic, this Louis XV style antique French marble mantel has a finely carved frieze decorated with shells and trails of flowers.

ABOVE *For a busy room with lots of other decorative features such as a patterned wallpaper or carpet, a fairly simply designed fireplace looks best. This corbelled marble mantel has a 'Queen Anne' fire grate with the hearth and inserts cut from Belgian fossil marble.*

LEFT *A period style fireplace in the 'country-house' style, this one has however been designed to fit the smaller proportions of a modern-day home. This simple but classic design is made entirely from reconstituted stone, though it looks exactly like the real thing.*

LEFT This stone fireplace looks like a 'dry stone' design. However, careful planning and lots of cement at the back of the slabs where it will not show, ensures that it is safe and ready for use.

BELOW LEFT A modern 'hole in the wall' gas flame-effect fire can look less than inspiring but not when it's set off on top of an almost black marble raised hearth. The effect is still strictly modern, but just a touch more stylish.

LEFT Your fireplace should match both the style of your room and its size. This scaled down antique Louis Phillipe marble surround is ideal for a small living room or bedroom.

BELOW RIGHT Reproduction wooden fire surrounds are now mass produced and available in all the home improvement shops, or you may be lucky enough to discover an original under layers of old paint, but the simplest way to give it a smart new look is with marble slips. You can buy these to fit standard openings – ask at your local fire shop – and fit them as a home improvement job.

RIGHT *Marble doesn't have to be used in large quantities to be effective. On this fireplace the white-painted wooden surround and gas flame-effect fire are shown off against a double layer of marble. The thinner top layer is cut from a white and pink veined marble, while the lower one has a dark grey pattern.*

ABOVE *This modern solid-fuel fire is set off handsomely against a grey veined black marble hearth and slips.*

LEFT *An attractive Louis XV antique marble fire surround, heavily carved and shaped.*

BELOW *A glorious rich pink-brown marble with black slate inserts makes a grand statement in a formal lounge.*

FLOORS

The saying goes, 'if you've got it, flaunt it', and when it comes to living room floors, stone really comes into its own. Without the hindrance of a bath, toilet or kitchen units to get in the way, you can really go to town with pattern and colour. Use marble mosaics or inlaid tiles to create borders around the edge of a room, to design a centrepiece or a 'rug' effect to highlight the seating area, the placing of a dining table or a walkway through the house.

On the other hand, plain stone floors can look just as beautiful in their simplicity, the perfect foil for other furnishings, with pastel coloured mats or brightly coloured oriental rugs as 'soft spots' beside chairs and sofas, or in front of a fireplace.

Practicalities

Care should be taken when choosing a stone floor as it is going to last a very long time. It could perhaps be with you forever, so it makes practical and economic sense to choose a colour or pattern you really like.

You should also take into consideration the floorcoverings laid in adjacent rooms for they should harmonize in both colour and style. If extra boards are going to be laid to level your floor, remember that with the stone on top this may well alter the

BELOW This intriguing 'rug' effect was formed using inlaid marble tiles set into a floor of co-ordinating plain marble tiles. Ready-to-lay, the tiles are individually cut and bonded together to create a huge number of designs. Work out ideas for a pattern such as this on squared paper, then sketch it on to a floor plan to see how it fits in with your furniture and seating.

RIGHT At first glance this intricate flooring looks like individual strips of marble, but it is in fact made up from square inlaid tiles laid in rows. A number of different patterns can be formed depending on how the tiles are put together.

height, perhaps creating a stepped effect at the threshold of rooms which may be unsightly and dangerous.

As stone – even fairly thin tiles – tends to be thicker than say a carpet or vinyl flooring, it's a safe bet that your existing doors will not close over the new floor. So you will have to take each one off, shave a slice off the bottom with a saw or plane, then re-hang it.

Marble and granite floors

Make the most of the different coloured stones available to create pattern to add interest to a large floor or to add detail in a smaller room.

ABOVE With the clever use of different colours and shades of marble, a clever 3-D effect can be created. This living room has in fact been laid using inlaid marble tiles which consist of precisely cut shapes stuck together. To use, they simply need to be bedded into a cement screed. Easy!

RIGHT The traditional black and white check floor is an all-time favourite which just won't date. In this long hallway large white slabs of marble alternate with quarter-sized black marble tiles to give this modern-day variation on a theme.

ABOVE *When walls and all other furnishings are plain a strongly patterned floor comes into its own. This detailed design from David Hicks International consists of three-colour triangles of marble, with alternate rows turned through 180° to form interlocking diamond shapes.*

RIGHT *Break up a long hallway with custom-made mosaic tiles. Made from tiny pieces of coloured marble and other coloured stones set into resin, the positions for these free-flowing leaves and flowers were carefully worked out so that each section stayed whole, while the co-ordinating plain marble tiles were cut to fit where necessary.*

RIGHT *In a modern apartment marble and granite take on a sleek style to blend perfectly with shiny mirror and ceramic wall decorations. This floor is made up of individual sections of stone, cut to echo the wall pattern.*

Stone and slate floors

Just like carpet and other flooring, plain stone looks best as a neutral base for other patterned or highly coloured furnishings. If you are likely to be redecorating in the years to come a plain coloured floor is also a safe bet to match with different colour schemes. A rough, plain stone floor can also suit different settings. While it looks perfectly at home in the country, it can also give a gentle rustic flavour to a town house or apartment.

TOP LEFT Every piece of fossilized stone takes on a slightly different tone and pattern, so careful planning will be needed to create an harmonious pattern between light and shade. Who says floor tiles have to be square? These unusually shaped hexagons fit in neatly with interlocking squares.

BOTTOM LEFT If you thought slate was black, think again, as it comes in a wide range of light and dark shades that can be used to create a truly stunning floor. These multicolour hexagonal tiles are predominantly black with a mixture of colours running through them from red to gold. This bold pattern works well with large spaces left between the tiles to take an ordinary cement filler.

RIGHT Hardwearing and easy to clean, softly coloured grey flagstones are just about the most practical flooring ever for a busy household.

LEFT *Square tiles allow you to design almost any pattern, so make full use of a range that includes slim border tiles too. These encaustic tiles are made in North Africa using a mixture of natural stone and powdered marble coloured with oxides which is still liquid while the design is formed, thus creating a wide choice of patterns. Being porous, the stone mellows once down – for instance, black goes grey which gives the floor an almost instant and attractive 'old' effect.*

LEFT *Even stone cut from the same quarry can differ in colour but by random positioning, it can be used to build a checker-effect floor. This unusual coloured African slate varies in colour from green to brown.*

ABOVE *French limestone flags are ideal for kitchens, hallways and conservatories. They can be bedded in with a wet cement mixture or loose laid on to sand.*

TOP LEFT *Use composition stones to make a dramatic show-piece floor. This particular range comes in a choice of seven designs in neutral and earthy colours, plus plain black and white tiles and a smaller border tile for a neat finish.*

TOP RIGHT *Highly practical, slate tiles are a versatile alternative to carpet in areas of heavy wear, such as hallways or changes in level. Like all types of stone, they blend easily into a natural setting, so it is simple to build the framework for a comfortable home. Here, roughly plastered walls and bleached wood provide the perfect foil for this rich, handsome stone.*

BOTTOM RIGHT *Grey limestone flags can look amazing when placed in a modern setting. This scheme relies on texture rather than colour: the rough riven stone, smooth satin finished wood and matt cream walls, all the perfect foil for one beautiful item of furniture.*

RIGHT *Encaustic tiles were introduced into medieval Spain by the Moors of North Africa. They fell from use until the Industrial Revolution but are once again popular today. The designs still show their African origins in the strong designs. This richly coloured Provencale tile and border is fairly typical – and the perfect way to brighten any dark or dreary floor.*

CHAPTER 6

EXTERIORS

The main consideration to be taken into account when choosing a stone for use on a patio or in a conservatory is that it is suitable for exterior use. It must be able to cope with extremes of temperature, from being baked in the summer sun to the frost and cold of the winter months of the year.

There are many types of reconstituted slabs, the most popular being a 'riven' effect, and compared with real stone they are much cheaper to buy. However there is a wide range of real stones that are suitable for exterior use too. Limestones are tough and hard-wearing, as are slate, quartzites and granites. So don't take a chance – the golden rule when choosing, is of course simply to ask if a particular stone is suitable.

Stone that is to be used inside a conservatory doesn't need to be quite as tough as that used outdoors, so all the stone grades that are suitable for general flooring will be alright here too. Your conservatory is sure to have a concrete base but do check that a damp proof membrane has also been included, then if necessary apply a self-levelling compound (see 'Kitchen Floors', page 24).

PATIOS

An area of paving in your garden adds colour, texture and pattern as well as providing an area for sitting, eating and enjoying the good weather.

To add interest to any paved area, try laying the stones in a pattern, with staggered joints brick-fashion, by varying large and small tiles or by changing the direction of pairs of oblong tiles to create a basket-weave design.

Good foundations

For stability, all exterior stonework needs a firm foundation or sub-base. This is usually made from a 10cm (4 inch) deep layer of hardcore with enough ballast to level the surface. However, this can be reduced where the soil is stable and for non-load bearing areas.

Other important points to bear in mind are:

DRAINAGE: The paving should slope away from the house so rainwater runs off. A fall of 1 in 100 lengthways is sufficient, and a fall of 1 in 40 across the width. On a small paved area this can be done by sloping the entire surface, but for a large patio it may be better to raise the centre slightly so the water runs off both edges.

DAMP-PROOFING: The top surface of the stone next to the house wall must be at least 15cm (6 inches) below the damp course. If for any reason this is not possible, then a gap of 75mm (3 inches) should be left between the wall and the edge of the stone.

OBSTACLES: Careful planning is needed for all stone patios but extra care should be taken if you have to deal with manhole covers, airbricks and drains. You may need to raise the level of the manhole cover or for

LEFT This African slate is very hard wearing, and with its unusual blue and gold colouring it makes an attractive focal point for the back of this house. Look for a range available in several different shapes and sizes to give plenty of scope for design ideas.

ABOVE *With wide spacings filled with a light coloured cement mortar, these beautiful African slate tiles offer a dramatic impact for a white-washed conservatory.*

a drain construct a low wall around it.

All exterior stones should be laid in a bed of sand and cement mortar. The usual mixture is one part cement to four parts sharp sand, but for spot bedding slabs or fixing individual slabs, you can use the bricklaying mix of one part cement to three parts sharp sand. For either mix, add a commercial plasticizer to make the mortar easier to work with and prevent it cracking as it drys.

Pile the dry ingredients together on a mixing board and mix thoroughly, then make a crater in the top and pour in a little water. Using a shovel, scoop the rim of the crater into the centre, then continue adding more water and turning the dry mixture until it is mixed into a stiff consistency that slides cleanly off your shovel.

Granite sets can either be laid in a bed of mortar or on a bed of sand. With the sand method, a vibrator

RIGHT *If you've got the space – and the cash – a specially commissioned mosaic has to be a talking point for any room in the house, let alone a conservatory. This stunning bird scene is made up from large square sections, using coloured marble and stones.*

must be used on the blocks after laying to compact the sand. You can hire these machines from a specialist hire company. Once completely level, make up a dry mixture of three parts sand to one part cement and spread it over the top of the stones, then using a stiff broom, sweep it into the joints. When the joints are full, damp down the surface with a fine spray from a watering can or hose. To dry the surface quickly, throw down some more fine sand on top and sweep it off. If the mixture settles unevenly in the joints, just repeat this operation.

When laying flagstones or large flooring tiles, use small offcuts of plywood about 1cm (½ inch) thick to act as spacers between them. Leave them in place until the mortar has dried, then fill the gaps as above.

RIGHT To give your garden patio a very different look, try using three differently coloured reconstructed stone shapes to create this convincing 3-D effect. Available in a choice of white, blue grey and grey , this range also comes in a number of other shapes and sizes.

BELOW Mosaic is the ultimate stonemason's art, a way of painting with stone, to create a floor or wall picture. This design was made by the English company Wessex Mosaics.

RIGHT Cast from natural reconstructed Portland stone, these simple ivory coloured slabs are inset with diamond shapes. Suitable for indoors and out, it is the ideal material to use both in a conservatory and to run out onto an adjacent patio.

CHAPTER 7

MATERIALS: MARBLE

The use of marble in interior design is widespread. Architects are extremely fond of using it to clad receptions and public buildings, but in the home it can be just as versatile, from cladding table and work tops to luxurious surface for use in bathrooms.

Strictly speaking marble is the geological name for limestones which have been completely re-crystallized by heat or pressure, but in general it is used as a name to describe any stone which can take a polish.

The main drawback when using marble is its porosity. Once in place it soaks up spilt liquids which can discolour the stone. So for this reason, although it will make an excellent kitchen work surface, it's important to realize that a lot of care will be needed to keep it looking in tip-top condition. However the stone can be treated to repel liquids, and where it can be used without worry it is a surprisingly trouble-free material.

It is found all over the world, though of course when one thinks of marble, the first country that springs to mind is Italy. Carrara is best known for its pure white marble that is used for sculpting. However the region around Carrara is still the largest marble producer in the world, and the marble produced there can be grouped:

- *Bianco Chiaro*, white with a few grey markings
- *Bianco Chiaro Venato*, white with stronger grey, and sometimes veined markings
- *Statuario*, divided into three groups of first statuary which is pure white, second statuary with some grey markings; and vein statuary
- *Bardiglio* which is a blue-grey.

Italy also produces coloured marbles including shades of cream, brown and red from Sicily and cream Botticino and Perlato from Sicily, Calacatta Rochetta (white with grey flecks), and Calacatta Vagli Rosato (pale pink to deep brown), both from Lucca.

Greece produces the classic Parian marble from the Island of Paros considered the finest white marble available; Skyros marble, white with orange to golden yellow markings; and Verde Antico from Larissa, a strong green.

Other main countries who export include Belgium (Belgian Black, Belgian Red), Spain (Red Bilbao), France (Napoleon Brown), Portugal (Rose Aurora), and Africa (Skyros, a white-pink background with thin red and black veins).

ONYX-MARBLE

Highly prized for its beautiful translucence, onyx-marble is a fine grained variety of calcite, thought to have been chemically deposited from standing sheets of water. Available in white, cream, yellow, green, red and brown, it comes from Algeria, Turkey, Iran, Pakistan and Mexico.

BIANCO NIPE, SPAIN

VENETO CROCICCHIO, ITALY

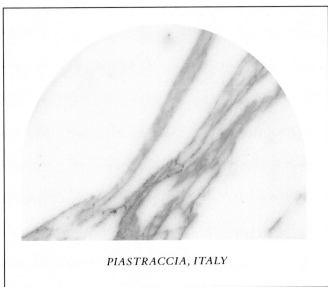

PIASTRACCIA, ITALY

ARNI ALTO, ITALY

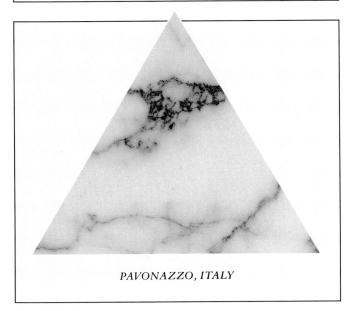

PAVONAZZO, ITALY

PAVONAZZO I, ITALY

BRECCIA VIOLETTA, ITALY

FIOR DEL POGGIO, ITALY

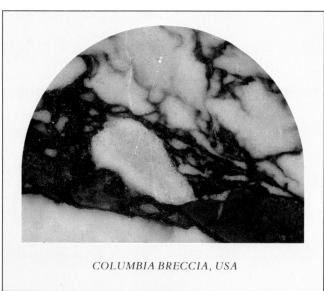

COLUMBIA BRECCIA, USA

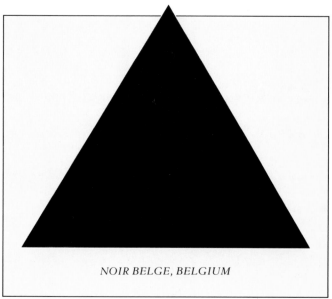

NOIR BELGE, BELGIUM

MOROCCAN BLACK, MOROCCO

PETIT GRANIT, BELGIUM

BLUE BELGE, BELGIUM

GRAND ANTIQUE DE PYRENEES, FRANCE

REGAL BLUE, USA

AZUL EXCURO, PORTUGAL

PORT ORO MACCHIE LARGA, ITALY

PORT ORO MACCHIE FINE, ITALY

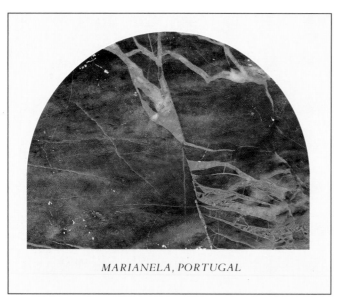

MARIANELA, PORTUGAL

ST ANNE MARBLE, BELGIUM

AZUL LAGOA, PORTUGAL

BARDIGLIO FIORITO, ITALY

BARDILLA BLEU FLEURI, ITALY

BRECHE ORIENTALE, FRANCE

BRECHE ORIENTALE, FRANCE

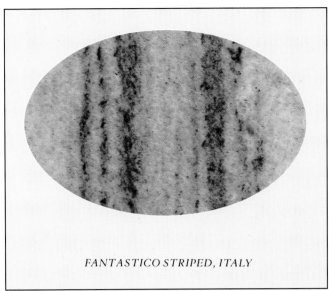

FANTASTICO STRIPED, ITALY

PIEDMONT BRECCIA, ITALY

BRONCEADO

AZUL DE SINTRA CLARO, PORTUGAL

BATESVILLE MARBLE, USA

NOTRE DAME RUBANE, FRANCE

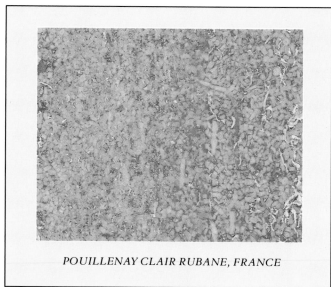

POUILLENAY CLAIR RUBANE, FRANCE

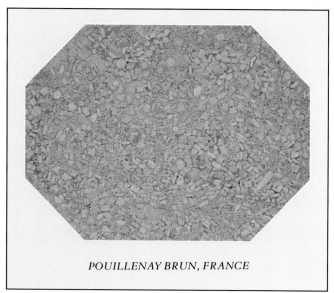

POUILLENAY BRUN, FRANCE

ROSE ST GEORGES, FRANCE

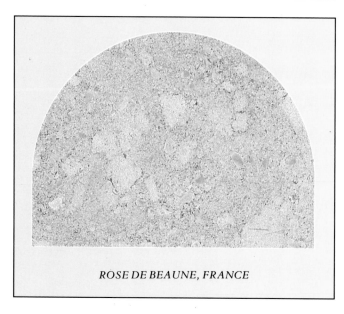

ROSE DE BEAUNE, FRANCE

CHAIMPO ROSICCIO, ITALY

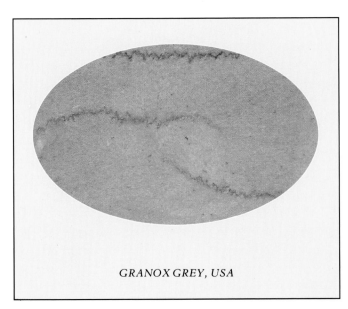

GRANOX GREY, USA

REPEN, YUGOSLAVIA

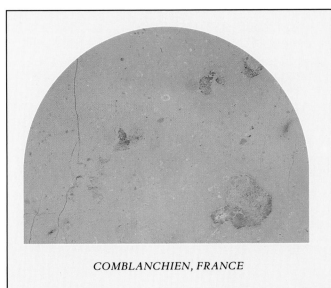

COMBLANCHIEN, FRANCE

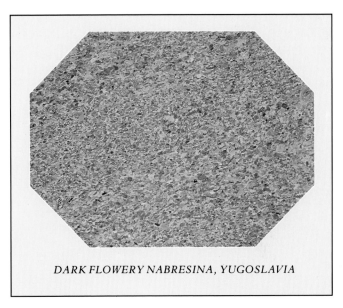

DARK FLOWERY NABRESINA, YUGOSLAVIA

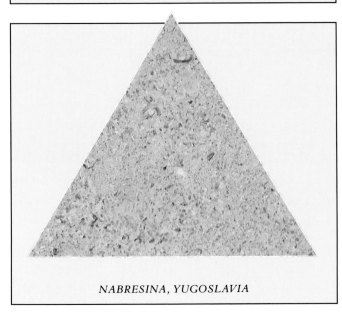

NABRESINA, YUGOSLAVIA

ROMAN STONE NABROSINA, YUGOSLAVIA

BIANCO DEL MARE, YUGOSLAVIA

LARRYS STONE, FRANCE

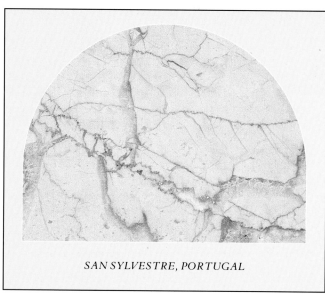

SAN SYLVESTRE, PORTUGAL

LIOZ ROSA, PORTUGAL

CIPOLLINO DORATO, ITALY

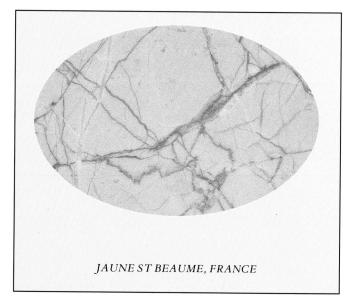

JAUNE ST BEAUME, FRANCE

CREMA ALICANTE, SPAIN

BRECCIATED SIENA, ITALY

SIENA GIALLO ANTICO MARBLE, ITALY

GIALLO SIENA, ITALY

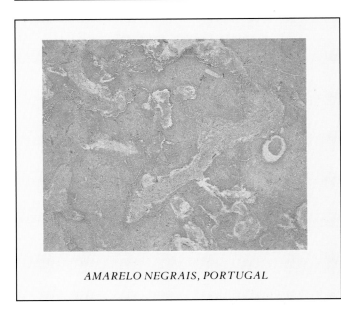

AMARELO NEGRAIS, PORTUGAL

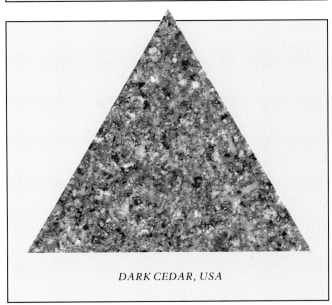

DARK CEDAR, USA

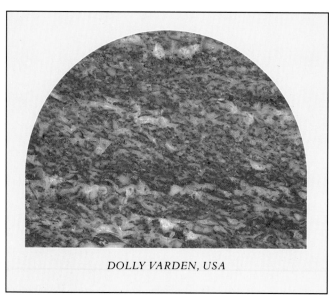

DOLLY VARDEN, USA

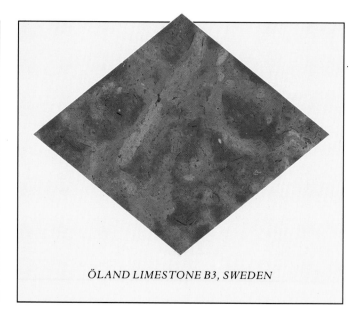

ÖLAND LIMESTONE B3, SWEDEN

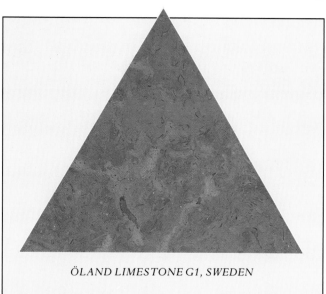

ÖLAND LIMESTONE G1, SWEDEN

BROCATELLO, SPAIN

GRIS DE VERDUN SAUMONÉ, BELGIUM

ST REMY GRIOTTE, BELGIUM

ST REMY BLEU, BELGIUM

ROUGE ROYALE DE SANTOUR, BELGIUM

ROUGE GRIOTTE DE SANTOUR, BELGIUM

PORFIRICO, ITALY

ÖLAND LIMESTONE B1, SWEDEN

ROSSO VERONA, ITALY

GRANITELLO ROSA MARBLE, SICILY

ROJO ALICANTE, SPAIN

VERMELHO, PORTUGAL

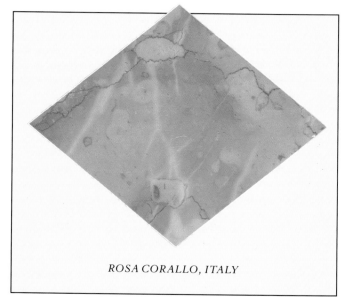

ROSA CORALLO, ITALY

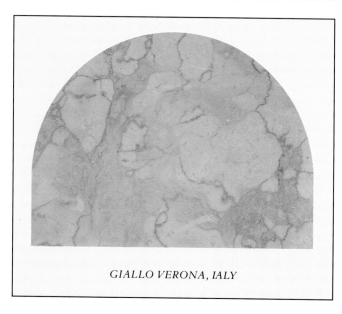

GIALLO VERONA, IALY

BRECCIA ROSA ARLECCHINO, ITALY

BRECHE ROSE, NORWAY

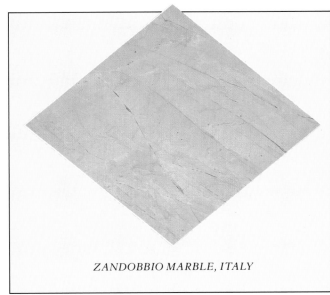

ZANDOBBIO MARBLE, ITALY

ROSA DEDO DE DAMA, PORTUGAL

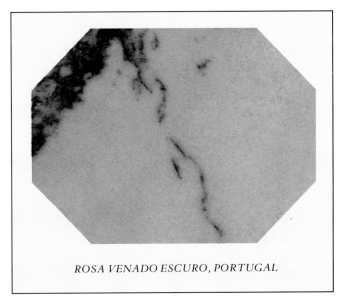

ROSA VENADO ESCURO, PORTUGAL

ALMISCADO, PORTUGAL

PORTUGUESE PINK, PORTUGAL

ROSSO DI LEVANTO, ITALY

VERDE TINOS, GREECE

VERDE DI GENOVA, ITALY

VERDE MODERNA, ITALY

VARALLO GREEN MARBLE, ITALY

VERDE DAMASCATO, ITALY

VERDE ANTICO DARK, GREECE

VERDE ANTICO DARK, GREECE

VERDE ANTICO PALE, GREECE

VERDE OLIVO, ITALY

GREEK CIPOLLINO

GREEK CIPOLLINO

ARNI ALTO, ITALY

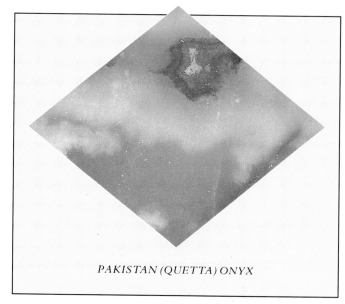

PAKISTAN (QUETTA) ONYX

ONYX MARBLE TURKEY

PORTORO MACCHIA FINE, ITALY

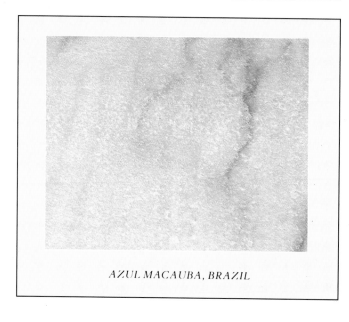

AZUL MACAUBA, BRAZIL

LASA MARBLE, ITALY

CHAPTER 8

MATERIALS: GRANITE

Formed from once molten magma, granite is the best known of all the igneous rocks, and it is widely used and found throughout the world in countries as far flung as the USA, Portugal, Italy and the UK, South Africa, India and Brazil.

It is without doubt the hardest and most durable natural stone used for building and because of its time-defying characteristics it is used extensively for external work. Largely unaffected by erosion and pollution, you'll see it cladding the outside of many important buildings and office blocks throughout the world.

Although the surface of granite can be finished in a number of ways it is only when the surface is polished that the true beauty of the stone is revealed. It is most often used to clad a feature wall, for fireplace surrounds, flooring, chimney breasts and most recently it has become popular as an almost indestructible kitchen working surface.

Happily, because granite is such a hard stone it can take the very roughest and toughest treatment without a care. To keep the polished surface looking good, just wipe over with a damp cloth and buff up with a clean dry cloth.

The main constituents of a true granite are quartz, alkali feldspar and mica and depending on the make-up of these minerals the granite can range in colour from the palest grey and pink to deep shades of red, though whatever the colour, the stone always shows the same distinctive coarse grain.

British granites include Shap which ranges from grey to a warm brown-red; Peterhead which is coarse-grained and a dark flesh colour; and Aberdeen granite with the world-famous Rubislaw quarry producing a dark blue-grey stone, while from the USA there is Imperial Mahogany, a red coloured stone with blue quartz and Texas Pink, which is pink to red in colour.

As well as the true granites, there are a number of other igneous rocks may also be grouped together and called granite, though their mineral make-up is slightly different.

Syenite is basically a granite without the quartz, made up mainly of the mineral feldspar, often with inter-grown crystals. Once polished, these crystals are known in the trade as 'butterfly wings'. It is most often sold as cut and polished slabs.

Granodiorite has a small or medium grain size and is light to dark grey in colour with black flecks.

Quartz Porphyry has large mineral crystals in a finer ground mass.

Basalt is perhaps the most common igneous rock found in the world, making up the Giant's Causeway in Northern Ireland with its amazing hexagonal pillars. It is a plain rock, grey to black in colour and mainly used in industry for road building and in the making of fibreglass.

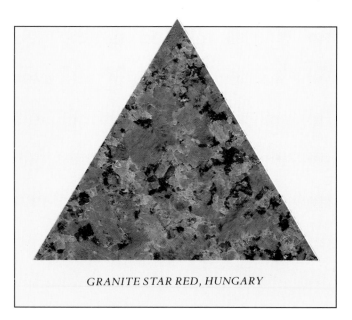

GRANITE STAR RED, HUNGARY

GRANITE BAGOTA RED, HUNGARY

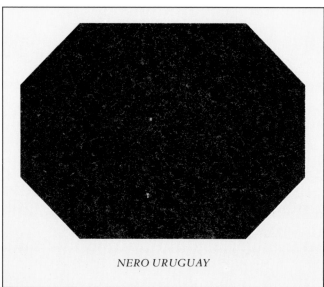

NERO URUGUAY

SHANDONG GRANITE, SWITZERLAND

PENMAENMAWR, WALES

TREVOR GRANITE, WALES

ROJO SIERRA CHICA GRANITE, ARGENTINA

ANDEER GRANITE, SWITZERLAND

SCANERA SHEARED GREEN GRANITE, SWITZERLAND

IMPERIAL RED GRANITE, SWEDEN

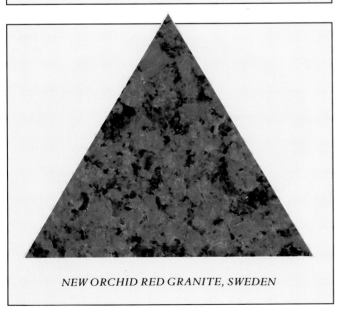

NEW ORCHID RED GRANITE, SWEDEN

GRANITE BLUEHILL, SWEDEN

IMPERIAL RED AKF GRANITE, SWEDEN

SWEDE ROSE ORIGINAL GRANITE, SWEDEN

RAMILO NO 6 GREEN GRANITE, SPAIN

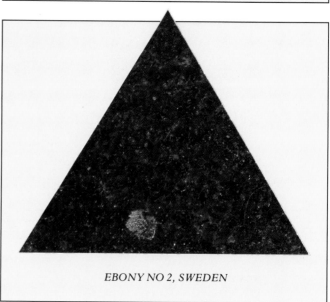

EBONY NO 2, SWEDEN

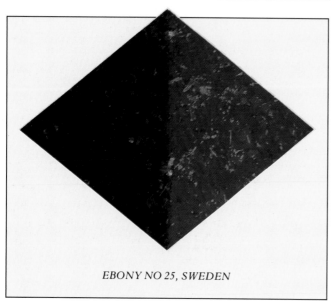

EBONY NO 25, SWEDEN

SOUTH AFRICAN BLACK

SOUTH AFRICAN GREY

SOUTH AFRICAN GREEN TWEED

PARYS GRANITE, SOUTH AFRICA

SOUTH AFRICAN GREY FLAME, TEXTURED

BRITZ BLUE GABBRO, SOUTH AFRICA

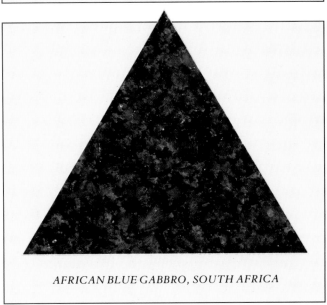

AFRICAN BLUE GABBRO, SOUTH AFRICA

CAK BLACK SWEDE, SWEDEN

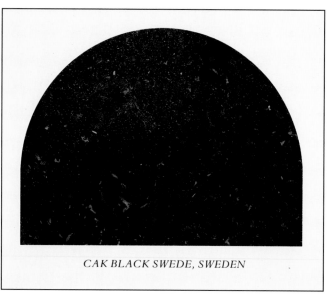

ROSSO SANTIAGO GRANITE, USSR

ROSSO ASSUNCION GRANITE, PORTUGAL

ROSA ARENA GRANITE, PORTUGAL

ST LOUIS SYENITE, PORTUGAL

STRZEBLOW GRANITE, POLAND

KOSMIN GRANITE, POLAND

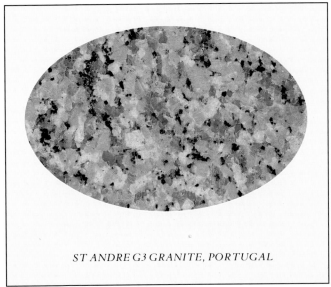

ST ANDRE G3 GRANITE, PORTUGAL

MICHALOWICE GRANITE, POLAND

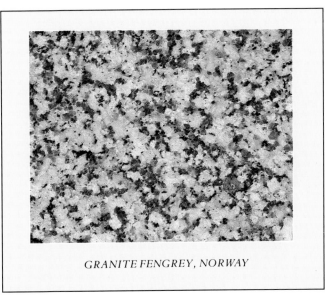

GRANITE FENGREY, NORWAY

BLUE PEARL SYENITE, NORWAY

GRANITO SAN LORENZO, SARDINIA

GRANITO ROSA ANTICO, SARDINIA

ATLANTIC GREEN SYENITE

VASSER PEARL

GRANITO BIANCO, SARDINIA

GRANITO ROSA LIMBARA, SARDINIA

RED STAR ORIENTAL

GRANITO ROSA, SARDINIA

BEOLA VERDE, ITALY

GHIANDONE, ITALY

PARADISO GRANITE, ITALY

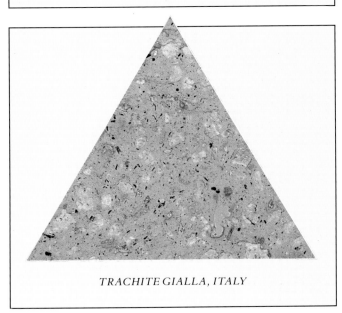

TRACHITE GIALLA, ITALY

TRACHITE GRIGIO CLASSICO, ITALY

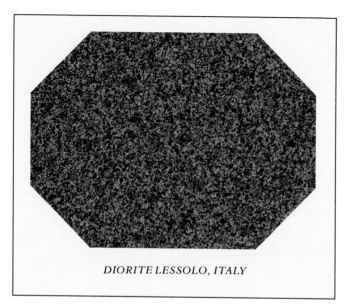

DIORITE LESSOLO, ITALY

SARIZZO GRIGIO, ITALY

BEOLA GRIGIA, ITALY

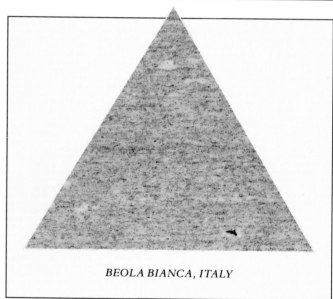

BEOLA BIANCA, ITALY

GHIANDONE DELLA VALMASINO, ITALY

GRANITO ST ELENA, ITALY

GRANITO ROSA DI BAVENO, ITALY

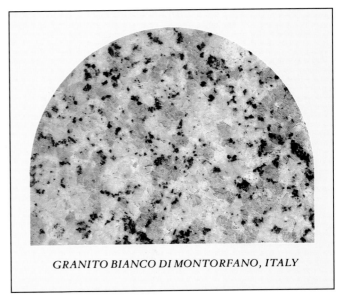

GRANITO BIANCO DI MONTORFANO, ITALY

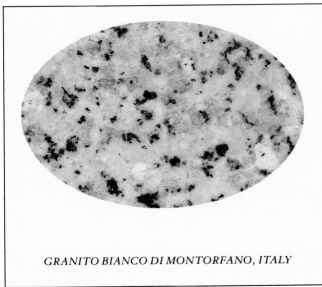

GRANITO BIANCO DI MONTORFANO, ITALY

SIENITE DI BALMA, ITALY

GRANITO VERDE VELMELENCO, ITALY

GRANITO VERDE DI MERGOZZO, ITALY

SARIZZO GHIANDONE GNEISS, ITALY

ROSSO GRANITE, INDIA

ROSSO MULTICOLOR GRANITE, INDIA

ROSSO MULTICOLOR RED FANTASIA

GREY FANTASIA GRANITE, INDIA

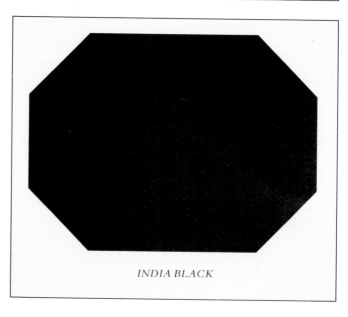

INDIA BLACK

INDIA GREEN

ROSSO RUBINO GRANITE, INDIA

CORAL RED GRANITE, INDIA

ROSSO NADO GRANITE, INDIA

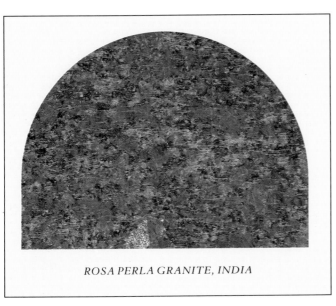

ROSA PERLA GRANITE, INDIA

LE HINGLE GRANITE, FRANCE

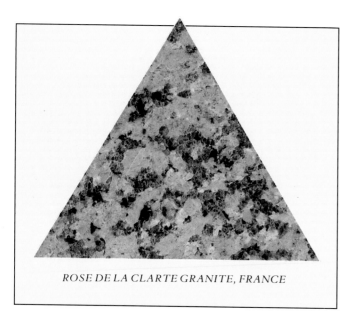

ROSE DE LA CLARTE GRANITE, FRANCE

TRAOUIEROS OR ROUGE TRAOUIEROS, FRANCE

MEGRITA GRANITE, FRANCE

BALTIC BROWN GRANITE, FINLAND

KARELIA RED GRANITE, FINLAND

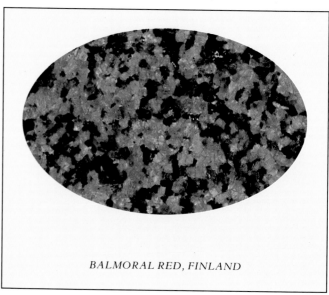

BALMORAL RED, FINLAND

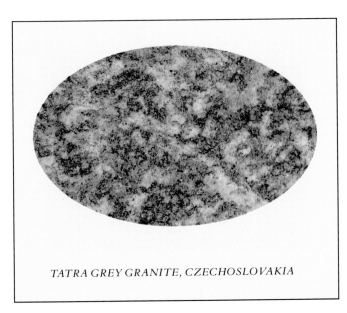

TATRA GREY GRANITE, CZECHOSLOVAKIA

AZUL BAHIA, BRAZIL

PEDRA DI GUARATIBA, BRAZIL

ROSSO BRAGANZA, BRAZIL

MARRON GUAIBA, BRAZIL

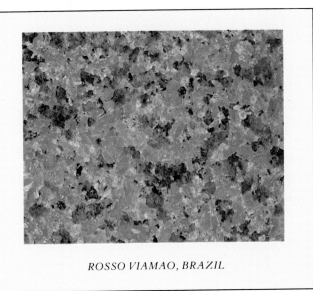

ROSSO VIAMAO, BRAZIL

ATIBAIA, BRAZIL

VERMENLO CAPAO BONITO MIADA, BRAZIL

PRETO TIJUCA, BRAZIL

EMATITA, ARGENTINA

AMARELO ABAETE, BRAZIL

ROSA BRITABA, BRAZIL

CHAPTER 9

MATERIALS: LIMESTONE

Limestone is a sedimentary rock, formed by sediments deposited on the beds of seas, rivers and lakes to settle into layered 'beds' or strata. Most limestone is used as a plain rock, but there are some types which can be polished and these are often grouped in under the heading 'marble'.

For interiors, these polished types are used for work surfaces, for flooring and for walls, while unpolished limestones are mainly used architecturally for pillars and to surround windows and doors, and for fire surrounds. Polished limestones should be treated just as for marble, while unpolished limestone should be cleaned quite simply by sponging off marks with warm water.

Limestones consist mainly of calcium carbonate, though most also contain some clay. As a general rule the lower the amount of clay the better the limestone will take a polish.

Limestone ranges in colour from almost white to soft pale browns, with textures from fine even grained stones to the smooth fossil-bearing types and coarse open textures. To classify these different types, limestone is grouped by origin: chemical, organic or clastic.

Chemical limestones such as travertine (see below) are formed by the evaporation of calcium carbonate from fresh or sea water.

Organic limestones consist partially or wholly of fossilised shells. Coral limestones are often called crinoidal.

Clastic limestone results from the erosion of existing limestone, consolidating into a new rock.

Countries who export polished limestones include Belgium with their famous 'Petit Granit', Germany's well known 'Solnhofen' which is full of beautiful fossils, as well as others from Israel, Portugal and France.

TRAVERTINE

Travertine is a chemical limestone. It was formed directly by the evaporation of calcium carbonate from water in rivers, and caves. Found mainly in Italy, there are large deposits in Tuscany, near Rome and in Tivoli. In the USA travertine is found at Dubois, Wyoming and in Nevada. All travertine is a light yellow or pink colour, and it used mainly for facing and paving.

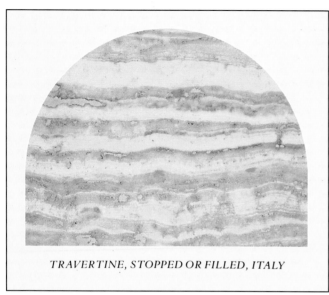

TRAVERTINE, STOPPED OR FILLED, ITALY

TRAVERTINO CHIARO CIPRESSO, ITALY

TRAVERTINO, ITALY

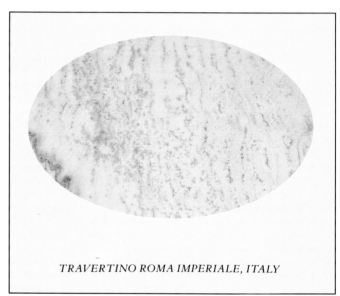

TRAVERTINO ROMA IMPERIALE, ITALY

TRAVERTINO SCURO ANTICO, ITALY

TRAVERTINO GIALLO, ITALY

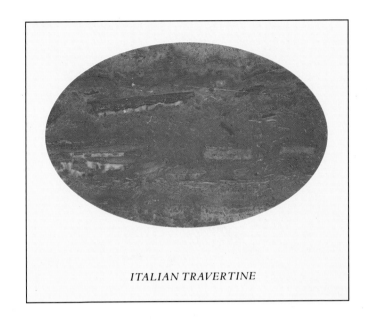

ITALIAN TRAVERTINE

TRAVERNIPE, SPAIN

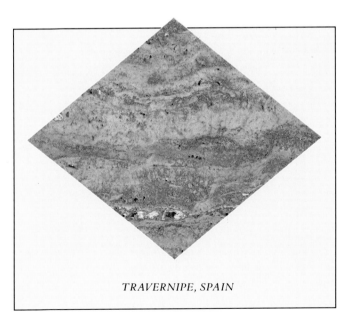

TRAVERNIPE, SPAIN

M A T E R I A L S : S A N D S T O N E

Sandstone is a clastic sedimentary rock which means that is has been formed by fragments of rock deposited either on the beds of rivers, seas or lakes. It has a layered structure which forms into 'beds' or strata.

Once quarried and sawn, sandstone usually has a sawn or clean rubbed finish and is most often used for walling, paving and monumental carvings. Domestically it is used for walls, fireplaces and in landscaping.

Sandstone is usually either pure white or honey coloured, depending on the other minerals present. The three minerals that form the actual sand grains are quartz, mica and feldspar. Mica has a tendency to congregate along certain planes and where this occurs, the rock will split easily and it is called a *flagstone*. Pure white sandstone is created from pure quartz grains, but this stone is rare; pale grey and cream sandstones are made up of a mixture of clear quartz with clear or white agrillaceous or calceous material, free from an iron stain; buff and yellow stones are usually coloured with limonite; pink, red and purple stones by haemitite; and dark grey and greenish stones with fragments of carboniferous material and chloride. Sandstone which has glauconite, a green mineral present is often called *greensand*. When feldspar is present the rock is called *arkose* and has a coarse, granular texture.

YORK PAVING, ENGLAND

ELLAND FLAGS, ENGLAND

ELLAND FLAGS, ENGLAND

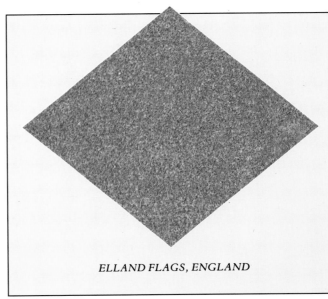

ELLAND FLAGS, ENGLAND

PENNANT SANDSTONE, WALES

PENNANT SANDSTONE, ENGLAND

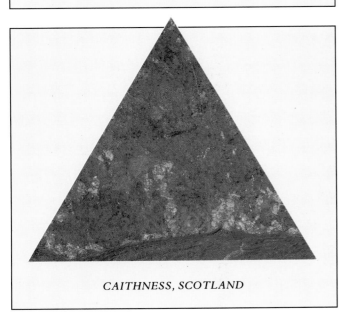

CAITHNESS, SCOTLAND

CAITHNESS, SCOTLAND

M A T E R I A L S : SLATE

The most obvious – and best known – characteristic of slate is its ability to split. This metamorphosed stone came about by clayey rocks being subjected to pressure over a long period of time. The flaky minerals re-orientated at right angles to the pressure and, therefore, a new grain was formed.

A fine grained rock, this ability to split readily and perfectly in constant direction, is called a slaty cleavage. To confuse the issue, there are a few sandstones and limestones which can also be split along their bedding planes into thin slabs, and in the past these have been used as roofing 'slates'. But they do not have a cleavage plane so cannot be classed as true slate. To set the record straight, a true slate has been formed from clays or volcanic ashes, deposited on a slowly subsiding sea floor. Covered with a thick layer of other deposits which compressed them into shale, earth movements , heat and pressure then caused both chemical and mechanical changes.

Because it can be cut so thinly, not surprisingly slate is a popular roofing material, but more unusually it can also be used as a work surface, for flooring, wall cladding, fireplaces and stair treads, and there is a long tradition of using slate to give a perfectly smooth, flat base for billiards tables.

Extremely durable, slate will not warp, twist or delaminate, and as a bonus is virtually maintenance free. When it is to be used as a table or work surface, the slate can be oiled to give it a soft polished effect. If it is to form kitchen work surfaces an appropriate vegetable or olive oil can be rubbed in, while for fireplaces and hearths any clear machine type of oil will do. The surface can then be re-oiled as and when necessary.

Found in the UK in Wales, Scotland, the Lake District and Cornwall, slate also comes from Spain, Portugal, Belgium and Italy.

BLAENAU FESTINIOG, WALES

PENRHYN BLUE, WALES

PENRHYN GREEN AND WRINKLED, WALES

GREY PENRHYN, WALES

GREY MOTTLED, WALES

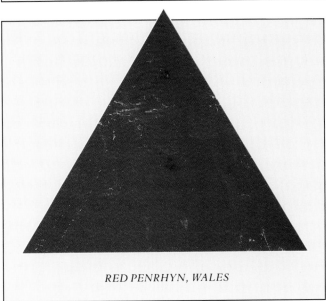

RED PENRHYN, WALES

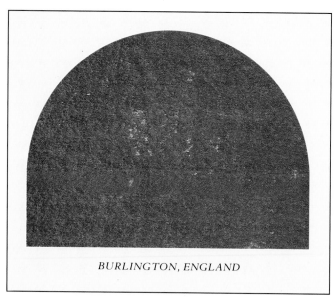

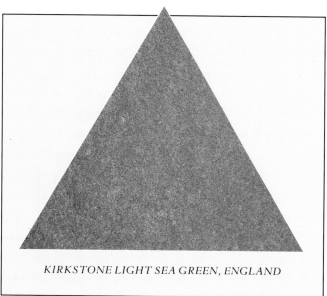

BURLINGTON, ENGLAND

DELABOL, ENGLAND

KIRKSTONE LIGHT SEA GREEN, ENGLAND

KIRKSTONE LIGHT SEA GREEN, ENGLAND

MANDALL'S SILVER GREEN GREY SLATE, ENGLAND

LIGHT SEA GREEN SLATE, ENGLAND

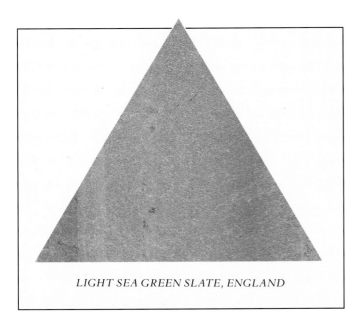

LIGHT SEA GREEN SLATE, ENGLAND

OLIVE GREEN SLATE, ENGLAND

NORWAY OTTA SLATE

NORWAY OTTA SLATE

MATERIALS: OTHER STONES

When searching for a particular shade or colour of stone you may also come across one or two other types which are used in the building trade but are not as widespread as those already mentioned.

Schist is the most abundant of all the metamorphic rocks including Alta quartzite (silver and grey-green) from Alta in Norway and Barge quartzite (olive, grey and light brown) from Mount Bracco in Italy. It is mainly used for the same applications as marble.

Gneiss has a coarse texture and is a banded rock made of layers of dark mica-schist and pale granite material often masquerading as a true granite. Switzerland produces a grey 'granite' and from Uruguay there is Ematita, a blue-grey to green 'granite', but both are Gneiss.

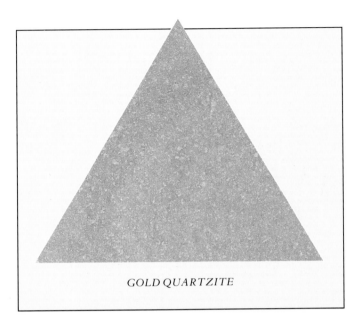

GOLD QUARTZITE

OLIVE QUARTZITE

OLIVE QUARTZITE

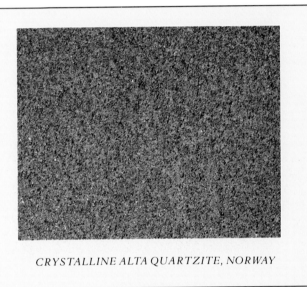

CRYSTALLINE ALTA QUARTZITE, NORWAY

GREY QUARTZITE

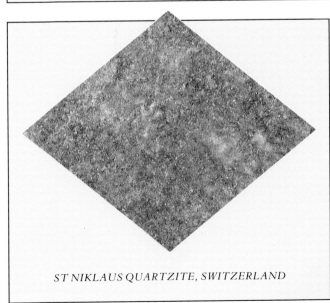

ST NIKLAUS QUARTZITE, SWITZERLAND

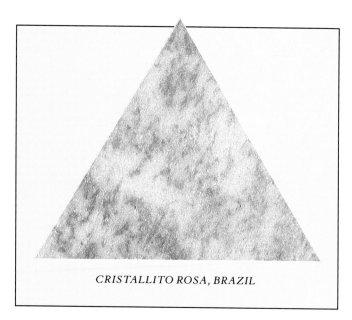

CRISTALLITO ROSA, BRAZIL

DIAMANT QUARTZITE, SOUTH AFRICA

SAFARI QUARTZITE, SOUTH AFRICA

ATLANTA IRISH SILVER GREEN

VAL D'AOSTA, ITALY

ALTAZITE NORWAY

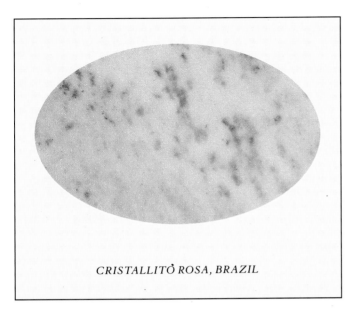

CRISTALLITÒ ROSA, BRAZIL

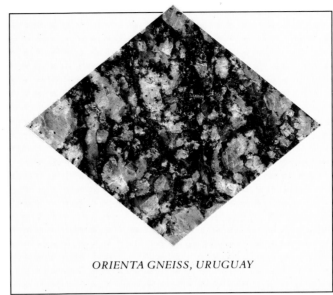

ORIENTA GNEISS, URUGUAY

SCANERA GREY GNEISS, SWITZERLAND

ALPSTONE GREY GNEISS, SWITZERLAND

GRIS MONTANA GNEISS, ITALY

SARIZZO CHIARO GNEISS, ITALY

ADDRESS BOOK

UNITED KINGDOM

Marble and Granite Suppliers

Adriatico Marble Ltd,
49 Kew Road,
Richmond,
Surrey TW9 2NQ
Tel: 081 948 5725
Fax: 081 940 5351

Anglo-Italian Marble Co Ltd,
181 Holloway Road,
London N7 8LX
Tel: 071 607 5139

Artmarbles Stone & Mosaic Co
Ltd,
Dawson Road,
Kingston-upon-Thames,
Surrey
Tel: 081 546 2023

W F Bannocks (Senr) Ltd,
Ambleside Marble Works,
1562 Stratford Road,
Birmingham B28 9HB
Tel: 021 745 1181
Fax: 021 745 6730

George V Beavin & Co Ltd,
4 Stanley Street,
Worksop,
Notts S81 7HX
Tel: 0909 473385

Blyth Marble Ltd,
Industrial Estate,
Lawn Road,
Carlton-in-Lindrick,
Worksop,
Notts S8l 9LB
Tel: 0909 730807

J Braithwaite & Sons Ltd,
Darwen,
Lancs
Tel: 0254 72516

Capital Marble,
Unit 10, Acklam Road,
London W10 5QZ
Tel: 081 968 5340
Fax: 081 968 8827

Chelsea Artisans PLC,
Unit 7,
The Ember Centre,
Lyon Road,
Hersham,
Surrey KT12 3PU
Tel: 0932 231000
Fax: 0932 241458
(Suppliers and fitters of 'Diamond
Marble' wall panels)

Classic Marble Co,
Horizon House,
Rainham Road,
Rainham,
Essex RM13 8SP
Tel: 04027 24141

Doncaster & Sons (Masons) Ltd,
l5 Micklegate,
York Y01 lJH
Tel: 0904 646864

Frame Sawn Supplies Ltd,
Crossgate Road,
Old Forge Industrial Estate,
Redditch,
Worcs B98 7SN
Tel: 0527 21626

Gormley (Marble Specialists) Ltd,
75 Kilburn Lane,
London Wl0 4AR
Tel: 081 960 8104

Walter W Jenkins & Co Ltd,
The Marbleworks,
Lymington Road,
Torquay,
Devon TQ1 4AS
Tel: 0803 39421

Imperial Facings Ltd,
Suite 200,
The Business Design Centre,
52 Upper Street,
Islington,
London N1 0QH
Tel: 071 288 6131

Leake's Masonry Ltd,
James Street,
Louth,
Lincs LN11 0LG
Tel: 0507 604828

Francis N Lowe,
New Road,
Middleton by Wirksworth,
Derbyshire
Tel: 062 982 22l6

Mactail Ltd,
Riverside Walk,
Lombard Wall,
Charlton,
London SE7 7SU
Tel: 081 858 5516

Marble Works Ltd,
33 Gunnersbury Lane,
Acton,
London W32 8ED
Tel: 081 992 1152

Midland Marble (Services) Ltd,
1558 Stratford Road,
Hall Green,
Birmingham B28 9HA
Tel: 021 744 3224
Fax: 021 745 2938

Natstone Sales Agency,
209 High Street,
Hampton Hill,
Middlesex TW12 1NR
Tel: 081 977 9094

Panceramic Ltd,
Eldon Way,
Crick,
Northants NN6 7SL
Tel: 0788 822129

C A Pisani & Co Ltd,
Transport Avenue,
Great West Road,
Brentford,
Middlesex TW8 9HF
Tel: 081 568 5001

Pure Marbles.
l58 Hagley Road,
Old Swinford,
Stourbridge,
West Midlands DY8 2JL
Tel: 0384 377676

Quality Marble Ltd,
Fountayne House,
Fountayne Road,
London N15 4QL
Tel: 081 808 1110
Fax: 081 885 2455

Thomas Group (Marble &
Granite) Ltd,
38 Higher Road.
Urmston,
Manchester M31 1AP
Tel: 061 748 6369
Fax: 061 746 8043

Whiteheads of London Ltd,
64 Kennington Oval,
London SE11 5SP
Tel: 071 735 1602

Stone Suppliers

Albion Stone Masonry
Ltd,
159-161 Boundary Road,
Merton,
London SW19 2DE
Tel: 081 543 7728

Baker Street Trading
Overseas Ltd,
332 Kilburn High Road,
London NW6 2QN
Tel: 071 624 0103

Bath & Portland Stone
Ltd,
Moor Park House,
Moor Green,
Corsham,
Wilts SN13 9SE
Tel: 0225 810456
Fax: 0225 811234

Burlington Slate Ltd,
Cavendish House,
Kirby-in-Furness,
Cumbria LA17 7UN
Tel: 022 989661
Fax: 022 989466

Cathedral Works
Organisation (Chichester)
Ltd,
Terminus Road,
Chichester,
West Sussex PO19 2TX
Tel: 0243 784225
Fax: 0243 771760

Cumbria Stone Quarries
Ltd,
Silver Street,
Crosby Ravensworth,
Via Penrith,
Cumbria CA10 3JA
Tel: 093 15227

Eternit TAC,
Meldreth,
Nr Royston,
Herts SG8 5RL
Tel: 0763 60421
Fax: 0763 62531

Granite Supply
Associations plc,
105 Urqhart Road,
Aberdeen AB2 1NH
Tel: 0224 636383

Allan Harris & Sons Ltd,
The Saw Mills,
Ashton,
Winkleigh,
Devon
Tel: 083 783 200

Jarret & Strachan,
3 The Cross,
St Michaels Road,
Minehead,
Somerset TA24 5JW
Tel: 0643 3381

G Kievel & Sons Ltd,
Lower Farm,
Ampfield,
Nr Romsey,
Hants SO5l 9BP
Tel: 0794 68865
Fax:0794 68914

Kirkstone Green Slate
Quarries Ltd,
Skelwith Bridge,
Ambleside,
Cumbria LA22 9NN
Tel: 05394 33296
Fax: 05394 34006

Stephen Lonsdale Ltd,
Building L,
Northfleet Industrial
Estate,
Lower Road,
Northfleet,
Kent DA11 9SN
Tel: 0322 847066

Minsterstone (Wharf
Lane) Ltd,
Ilminster,
Somerset TA19 9AS
Tel: 0460 52277
Fax: 0460 57865

Natural Stone Carpet Ltd,
Unit 13,
Burfield Park,
South Road,
Hailsham,
East Sussex BN27 3JL
Tel: 0323 843068
Fax: 0323 410416

Natural Stone Products
Ltd,
Bellmoor,
Retford,
Notts DN22 8SG
Tel: 0777 708771

Naturestone,
1a King's Road Park,
King's Ride,
Ascot,
Berks SL5 8AR
Tel: 0990 27617
Fax: 0990 21462

Scottish Natural Stones
Ltd,
Edinburgh Road,
Springhill,
Shotts ML7 5DT
Tel: 0501 23248
Fax: 0501 23058

Flooring Tiles

Cardiff Mosaic & Terrazzo Co
Ltd,
Sloper Road,
Cardiff,
S. Glamorgan CF1 8AB
Tel: 0222 21766

Domus Tiles Ltd,
33 Parkgate Road,
London SW11 4NP
Tel: 071 223 5555

Fired Earth,
Middle Aston,
Oxfordshire OX5 3PX
Tel: 0869 40724
Fax: 0295 810832

Holley, Hextall & Associates,
Chittor,
Chippenham,
Wilts SN15 2EL
Tel: 0380 850039

Marble Flooring Specialists Ltd,
110 Ashley Down Road,
Bristol BS7 9JR
Tel: 0272 420221
Fax: 0272 41969

Marble World,
8 Commercial Way,
Abbey Road,
Park Royal,
London NW10 7XF
Tel: 081 965 0014

Paris Ceramics,
543 Battersea Park Road,
London SW11 3BL
Tel: 071 228 5785
Fax: 071 924 2282

A Quiligotti & Co Ltd,
Newby Road,
Hazel Grove,
Stockport.
Cheshire SK7 5DR
Tel: 061 483 1451
Fax: 061 456 0209

Reed Harris Ltd,
Riverside House,
Carnwath Road,
London SW6 3HS
Tel: 071 736 7511

Shops and Fitters

JCD London Ltd,
13 New Burlington Street,
London W1X 1FR
Tel: 071 734 3507
Fax: 071 734 0873
(Specialist bathroom suppliers)

Mander & Germain Ltd,
60 Kimber Road,
London SW18 4PP
Tel: 081 871 1326

Morris Marble Works,
Market Lane,
Whickham,
Newcastle-upon-Tyne NA16 4TH
Tel: 091 488 7289

Wessex Mosaics,
Bretworth,
Woodlands Road,
Portishead,
Bristol BS20 9HE
Tel: 0272 843022

Zarka Marble Ltd,
41a Belsize Lane,
London NW3 5AU
Tel: 071 431 3042

Kitchen Worksurfaces

John Lewis of Hungerford,
Unit 2,
Limborough Road,
Wantage.
Oxon OX12 9AS
Tel: 02357 68868

Siematic UK Ltd,
11-17 Fowler Road,
Hainault Industrial Estate,
Ilford,
Essex IG6 3UU
Tel: 081 500 1944

Fireplaces

Bell Fireplaces,
A Bell & Co Ltd,
Kingsthorpe Road,
Kingsthorpe,
Northampton NN2 6LT
Tel: 0604 712505

Ideal Fireplaces,
300 Upper Richmond Road West,
East Sheen,
London SW14
Tel: 081 878 7887

JLB Group Ltd,
Cozyfire Court,
Boundary Lane,
Liverpool L6 5JG
Tel: 051 263 9545

Marble Hill Fireplaces Ltd,
72 Richmond Road,
Twickenham,
Middx TW1 3BE
Tel: 081 892 1488

Minsterstone (Wharf Lane) Ltd,
Ilminster,
Somerset TA19 9AS
Tel: 0460 52277
Fax: 0460 57865

Petit Roque Ltd,
5a New Road,
Croxley Green,
Nr. Rickmansworth,
Herts WD3 3EJ
Tel: 0923 779291

Wickes Building Supplies Ltd,
Wickes House,
120-138 Station Road,
Harrow,
Middx HA1 2QB
Tel: 081 863 5696

Restoration, Conservation & Cleaning

A. Bell & Co Ltd,
Kingsthorpe Road,
Kingsthorpe,
Northampton NN2 6LT
Tel: 0604 712505
(makers and suppliers of cleaning
materials, polish and sealers for
marble, slate and stone)

Beech Restoration Ltd,
The Old Coach House,
2 Beech Avenue,
Sherwood Rise,
Nottingham NG7 7LL
Tel: 0632 609237

New Stone & Restoration Ltd,
1 Pembroke Road,
Ruislip,
Middlesex
Tel: 0895 676184
Fax: 0895 621574

Nigel Cox Ltd,
128-130 Putney Bridge Road,
London SW15 2NQ
Tel: 081 871 2755

Rattee and Kent,
Purbeck Road,
Cambridge,
Cambs CB2 2PG
Tel: 0223 248061

Information

Association of Flooring
Contractors,
23 Chippingham Mews,
London W9 2AN
Tel: 071 286 4499

The Building Centre,
26 Store Street,
London WC1E 7BT
Tel: 071 637 8361

Crafts Council,
12 Waterloo Place,
London SW1Y 4AU
Tel: 071 930 4811
(The aim of the Council is to help
Britain's craftsmen to improve
their standards and sell their
work.)

Geological Museum,
Exhibition Road,
South Kensington,
London SW1
Tel: 071 938 8765
(Houses the national collection of
building and decorative stones.)

Granite Guild,
170 High Street,
Lewes,
East Sussex BN7 1YE
Tel: 07916 77374

Guild of Master Craftsmen,
170 High Street,
Lewes,
East Sussex BN7 1YE
Tel: 07916 77374
(Publishes an annual directory of
all its members, all skilled and
practising a craft, trade or art.)

National Fireplace Manufacturers
Association,
PO Box 35 (Stoke),
Stoke-on-Trent ST4 7NU
Tel: 0782 44311
(Membership includes firms who
manufacture and install stone,
marble and slate fireplaces.)

National Master Tile Fixers
Association,
c/o Joliffe Cork & Co,
Elvian House,
18-20 St. Andrew Street,
London EC4 3AE
Tel: 071 628 6441

National Stone Centre,
Ravenstor Road,
Wirksworth,
Derbyshire DE4 4FR
Tel: 0629 824833
(An educational, exhibition and
trade centre with training and
advisory services
to promote a better understanding
of the quarry and stone industry.)

Royal Institute of British
Architects,
66 Portland Place,
London WlN 4AD
Tel: 071 580 5533

Stone Federation,
82 New Cavendish Street,
London WlM 8AD
Tel: 071 580 5588
(The prinicipal body representing
the stone industry in the UK.)

AUSTRALIA

Gosford Quarries Pty Ltd,
PO Box 86,
300 Johnston,
Annandale,
NSW 2038,
Tel: (02) 827 2555. 82 1818
(Suppliers of sandstone)

Master Stone Masons Association
(Victoria),
Ground Level,
l8-20 Lincoln Square North,
Carlton,
Vic 3053
Tel: (03) 347 2855

Melocco Pty Ltd,
l00-110 Euston Road,
Alexandria
NSW 2015
Tel: 5199333
(Quarry owners, wholesalers and
importers, supply and fix)

Stone & Terrazzo Association of
NSW,
60 York Street,
Sydney,
NSW 2000
Tel: 290 0700

UNITED STATES

Information

American Granite Assn,
933 High Street,
Suite 220,
Worthington.
OH 43085
Tel: (614) 885-2713
(Part of the American Monument
Assn, they publish 'Stone in
America' magazine.)
Assn of Tile, Terrazzo, Marble
Contractors and Affiliates,
PO Box 13629,
Jackson,
MS 39236
Tel: (601) 9392071
(Promotes the use of tile, terrazzo
and marble.)

Barre Granite Association,
PO Box 481,
Barre,
VT 0554l
Tel: (802) 476 4131

Building Stone Institute,
420 Lexington Avenue,
New York,
NY 10170
Tel: (212) 490 2530
(Gives information to architects,
decorators etc.)

Cultured Marble Institute,
435 N Michigan Avenue,
Suite 1717,
Chicago,
IL 60611
Tel: (312) 644-0828)
(Promotes firms who make
cultured marble products.)

Elberton Granite Association,
PO Box 640,
Elberton,
GA 30635
Tel: (404) 283 2551

Indiana Limestone Institute of
America,
Suite 400,
Stone City Bank Building,
Bedford,
Indiana 47421
Tel: (812) 275 4426

Italian Trade Association,
499 Park Avenue,
New York,
NY 10022
Tel: (212) 980-1500
(Promotes imports of Italian
products and publishes a marble
newsletter.)

Marble Collectors Unlimited,
Box 206,
Northboro,
MA 01532
Tel: (617) 393-2923
(Club for marble collectors and
enthusiasts.)

Marble Institute of America Inc,
33505 State Street,
Farmington,
Michegan 48024
Tel: (313) 476 5558

National Building Granite
Quarries Assn,
PO Box 482,
Barre,
VT 05640
Tel: (802) 476-3115

National Quartz Producers
Council,
PO Box 1719,
Wheat Ridge,
CO 80034
Tel: (303) 424-6722

National Stone Assn,
1415 Elliot Place, NW,
Washington DC 20007
Tel: (202) 342-1100

Granite Suppliers

AAA Hegarthy & Sons,
295-E Skyline Dr,
Easton,
PA 18042
Tel: (215) 2528l00

American Granite Quarries Inc,
PO Box 960,
Elberton,
GA 30635
Tel:(404) 283-2613

Arkansas Missouri Granite Works,
Box 335E,
Imboden,
AR 72434
Tel: (501) 869 286l

Comet Int Co Inc,
PO Box 814809-E,
Dallas,
TX 7538l
Tel: (214) 620-1888

The Evanswinn Co,
PO Box 658,
Elberton,
GA 30635
Tel: (404) 283 8821

Field Granite Int Ltd,
3434-E Heritage Dr,
Minneapolis,
MN 55435-2262
Tel: (612) 920-9145

Rock of Ages Corp,
PO Box 482-E,
Barre,
VT 05641
Tel: (802) 476 3115

Cultured Marble Suppliers

Esquire Marble Company,
15104E Sardis Road,
Mabelvale,
AR 72103
Tel: (501) 847-2797

Gesmar Marble Corp,
1806 Summit Avenue,
Richmond,
VA 23230
Tel:(804) 358-4956

Wildon Industries Inc,
Rte 512,
PO Box 176,
Mt Bethel,
PA 18343
Tel: (215) 588-1212

Marble Suppliers

D & B Tile Distributors,
5800 Rodman Street,
Hollywood,
FL 33023
Tel: (305) 983-6373

Georgia Marble Co,
3460 Cumberland Pky,
NW Atlanta,
GA 30399
Tel: (404) 432-0131

London Universal Trading Co,
3658E Caminito Carmel Landing,
San Diego,
CA 92130
Tel: (619) 755-8383

Loughman,
2121 Walton Road,
St. Louis,
MO 63114
Tel: (314) 426-0321

McBride Stone Quarries,
Box 220 AE-Rt 4,
Batesville
AR 72501
Tel: (501) 793 7285

Monarch Metal Products Corp,
PO Box 42381-E,
Cincinnati,
OH 45242-0381
Tel: (513) 791-0595
(Bathroom sink tops with integral
bowls.)

Moretti-Harrah Marble Co,
PO Box 330-E,
Sylacauga,
AL 35150
Tel: (205) 249-4901

New York Marble Works Inc,
1399 Park Avenue,
New York,
NY 10029
Tel: (212) 534-2242
(Fireplaces, book-ends, sink tops,
tables etc.)

Stone International Inc,
788 Scottdale road,
PO Box 33569,
Decatur,
GA 30033-0569
Tel: (404) 292-0135

Vermont Marble Co,
6l Main Street,
Proctor,
VT 05765
Tel: (802) 459-3311

GLOSSARY

adit: a sloping tunnel or shaft driven through a hill or mountain side to reach beds of rock

agrillaceous: a fine grained sedimentary rock with grains less than 1/16mm, eg clay

ashlar: a block of stone with straight edges for use in building

bed: a layer of sedimentary rock

bedding plane: surface in sedimentary rock parallel to the original surface on which the sediment was deposited

breccia: a clastic sedimentary rock with angular fragments

clastic: sediments formed from the breaking up of earlier rocks

cleavage: the tendency of some rocks to split or break along smooth planes that are more or less parallel

columnar jointing: in igneous rocks, a regular six-sided form of jointing that produces regularly shaped pillars or columns

conglomerate: a rock composed of rounded fragments, anything from a few millimetres to several centimetres in diameter.

country-rock: the rock or rocks into which an igneous intrusion is placed

crystallized: to form into crystals

dyke: an igneous intrusion, rather like a wall, into the surrounding rocks

extrusive rock: an igneous rock formed by the cooling of magma on the earth's surface

fault: a fracture within a rock mass where the rocks on one side have moved in relation to those on the other side

fossiloferous: a rock containing fossils

grains: the individual mineral pieces or crystals that make up a rock

groundmass: the main part of an igneous rock made up of finer grains in which the larger crystals are set

heterogeneous: formed from several types of material

homogeneous: formed from just one material

igneous: rocks which have solidified from a molten state

intrusive: igneous rock formed by the cooling of magma inside the earth's crust

magma: liquid or molten rock material, it is called lava when it reaches the earth's surface

metamorphic: rocks which changed from another rock by the action of either heat, pressure, or both

mosaic: small pieces of marble or other stone inlaid into a design or geometric pattern

oolite: the small round particles which make up a sedimentary rock. On mass they look just like fish eggs

opencast: the method of mining near the surface by cutting into it from above rather than digging underground

overburden: the unusable rock and matter lying over the stone to be quarried

petrology: the study or rocks, their origin and what they are made from

plutonic: igneous rocks which have formed from magma at a great depth in the earth's crust

polychromatic: having various or changing colours

quartzite: the metamorphic equivalent of a quartz sandstone which has re-crystallised into closely fitting granules.

rock: any natural material formed of a single mineral or various minerals

scagliola: a material developed in the 17th century in Northern Italy to duplicate marble. It is made from coloured plaster and isinglass with inset marble chips. It can be polished to give a gloss finish

sedimentary: rocks formed by the transformation of existing rocks by gravity, atmosphere and living organisms

sill: an igneous intrusion that is more or less horizontal but forms into a single step shape

statuary marbles: those marbles used for sculpture

steeling: mending a vent (hairline) crack by cutting grooves on the reverse side of a slab of stone and inserting strips of metal

strata: layers or beds of sedimentary rock

structure: the overall character of a rock

terrazzo: a floor or wall finish made by setting marble or other stone chips into a layer of mortar and polishing the surface

texture: the total characteristics of a rock given by the size and shape of its grains

venting: a natural hairline crack in the stone

INDEX

PICTURE CREDITS

p6 Elizabeth Whiting Associates; p8 Zeyko Kitchens; p12/15 Elizabeth Whiting Associates; p16 Zeyko Kitchens; p17 Mark Wilkinson Furniture; p18 top Siematic Kitchens; bottom John Lewis of Hungerford; p19 Naturestone; p20 John Lewis of Hungerford; p21 Siematic Kitchens; p22/23 Smallbone Kitchens; p24/25 Naturestone; p26 Bisque Radiators; p27 Paris Ceramics; p28 Bisque Radiators; p29 Gerhardt Kievel & Sons Ltd; p30 Elizabeth Whiting Associates; p31 all Siematic Kitchens; p32 Smallbone Kitchens; p34 John Lewis of Hungerford; p35 Mark Wilkinson Furniture; p36 Elizabeth Whiting Associates; p37 Minsterstone; p38/39 all JCD; p40 Twyford Bathrooms; p41 Showerlux; p42–45 Elizabeth Whiting Associates; p46–49 David Hicks International; p50 Naturestone; p51 JCD; p52 Marble Hill Fireplaces; p53 top; Chimneypieces; bottom Minsterstone; p54 Elizabeth Whiting Associates; p55 top Chimneypieces; bottom left Jetmaster Fires; bottom right Wickes DIY stores; p56 top Jetul Fires; bottom Jetmaster Fires; p577 top Chimneypieces; bottom Jetmaster Fires; p58–60 Domus Tiles Limited; p61 David Hicks International; p62 top David Hicks International; bottom Wessex Mosaics; p63 David Hicks International; p64 top Gerhardt Kieval & Sons Ltd; bottom Fired Earth; p65 Paris Ceramics; p66 Fired Earth; p67 top Paris Cermaics; bottom Naturestone; p68 all Paris Ceramics; p69 top Elizabeth Whiting Associates; bottom Fired Earth; p70 Naturestone; p72 Fired Earth; p73 Wessex Mosaics; p74 top Minsterstone; bottom Wessex Mosaics; p75 Minsterstone; p77–121 The Natural History Museum, London.